THE LAST PARADISE

THE LAST PARADISE

An American's 'discovery'
of Bali in the 1920s

BY
HICKMAN POWELL

ILLUSTRATIONS BY
ALEXANDER KING

PHOTOGRAPHS BY
ANDRE ROOSEVELT

SINGAPORE OXFORD NEW YORK
OXFORD UNIVERSITY PRESS

Oxford University Press

Oxford   New York   Toronto
Petaling Jaya   Singapore   Hong Kong   Tokyo
Delhi   Bombay   Calcutta   Madras   Karachi
Nairobi   Dar es Salaam   Cape Town
Melbourne   Auckland

and associates in
Beirut   Berlin   Ibadan   Nicosia

OXFORD is a trademark of Oxford University Press

First published by Jonathan Cape and
Harrison Smith Inc., New York in 1930
First issued in Oxford in Asia Paperbacks 1982
Second impression 1986

Reissued as an Oxford University Press paperback 1986

ISBN 0 19 582537 3

Printed in Malaysia by Peter Chong Printers Sdn. Bhd.
Published by Oxford University Press Pte. Ltd.,
Unit 221, Ubi Avenue 4, Singapore 1440

FOR G. W.

CONTENTS

INTRODUCTION

For five years I have lived a happy and contented life in Bali.

While there, by the greatest of good luck, I met Hickman Powell. He will tell you how I insisted that he write this book, but being a very modest young man he will not mention the reasons why I picked him out for this very difficult task.

People are kind enough to admire my pictures, both still and movies; audiences seem to enjoy my lectures and seldom leave before the end, but I have yet to find anybody who has ventured to assert that I am a successful writer. As an author I must confess that I am a most dismal failure.

There I was in Bali—"The Last Paradise"—filled with knowledge about the people, their interesting customs and their religion; before me I saw taking place the slow but sure westernization of this glorious country, and I was unable to make a record of what I had seen and still saw. Somebody had to do it.

After an hour or two spent with Powell, I realized that this keen young American newspaper man had assimilated in a few hours infinitely more than the average person would have in the same number of months. He had that peculiar knack which is characteristic of the good newspaper reporter of digging deep below the surface and delving into the psychology of things unseen; he was also avid for facts, exact facts, and added to this, he had the soul of an artist.

Powell had been struck by the culture of the Balinese; their religion fascinated him; he had been charmed by their wonderful music, and had spent hours admiring their dances. When I discovered that he could write, I felt that he was just the man I had been looking for. He was ambitious. When I suggested he remain in Bali to write the first book in English about the island and its people, he could not resist the temptation.

Powell has got the spirit of Bali. His sketches are true to life. He has dug deeply into the souls of the people, and tells charmingly of what he found or did not find there. He now loves them as I do, because he understands them as I do.

You will be told that the Balinese are the greatest artists of this age, and still more, that every Balinese, man or woman, is an artist.

A few days ago I was the guest of a very charming couple in Denver, Colorado. I was dressing preparatory to delivering a lecture when I suddenly saw a great light. I had struck upon the explanation of the phenomena; I had found an answer to the question—why are all Balinese artists?

We know that all great periods of art occur only when certain specific conditions exist. Egypt, during the Fourth and Eighteenth dynasties; the Chinese Tang and Sung dynasties; Greece, Rome, and much later, in our case, the Renaissance, also that of Japan, took place because these respective states were, first of all, rich and prosperous at the time. Trade and commerce were good. The great lords and wealthy merchants found time to encourage the arts, and though these wealthy people had leisure, they had not the creative instinct. These great lords were not artists themselves, but their affluence enabled them to foster the arts and encourage the artists financially, keeping them employed for long periods of time. They were the patrons of the arts of their time.

The two conditions are, therefore, wealth and leisure; besides these two there must be an objective, which was generally found in the religion prevalent during these various periods.

Now let us see what conditions existed and still exist in Bali. The Balinese are a rich nation, one of the wealthiest per capita in the world, and this wealth is very evenly divided among the million inhabitants of the island. Condition One is fulfilled—wealth.

Due to the richness of the soil, the Balinese, by working four months a year, can produce all he needs for the wants of himself and family; the rest of the time is his own. Condition Two exists—Leisure.

Finally, his very life centers around his religion, a beautiful mixture of Animism and Hinduism. Here is the needed incentive.

For centuries these same conditions have existed. The people, therefore, turned to art as a pastime. Whether sculpture, painting, music, or dancing, they simply had to produce or to give breath to all these, since they could not help themselves.

Now, this nation of artists is faced with the Western invasion, and I cannot stand idly by and watch their destruction.

Nowhere in the world has the aborigine been able to resist the invasion coming from the West. The age of steel, as typified by this country, has crept into the lazy, happy, contented East, leaving behind a trail of unhappiness and sorrow. Can Bali, which stands in a peculiar geographical and economic position, be able to resist and retain its individuality?

Bali is self-contained. The million inhabitants live on what they raise themselves and do not have to import food or other commodities. There are no mines; there are no industries; there is nothing that the white man can acquire except the fertile lands which are and will permanently remain in the hands of the natives, thanks to the intelligent policy of the Dutch Government, which has taken as its slogan, "Bali for the Balinese." No white man or Asiatic can acquire an acre of land except for residential purposes. I have often been asked whether the modern methods of cultivation would increase the yields in Bali. I always answer that we have nothing to teach these wonderful little farmers whose crops are far beyond anything that we have even dreamt of in this country. May I give an example?

In my youth I raised rice in Texas. My average crop, per acre, was 10 sacks, each weighing from 170 to

180 pounds. In Bali the grand average is over 21 sacks, weighing from 190 to 200 pounds, and they can raise three crops a year as against one in Texas. At the best, by the building of dams in the higher regions, the acreage could be increased about 10%. I mention this simply to show that we can do little or nothing to increase the wealth of the Balinese.

The exports of pigs, cattle, copra, and coffee have, in the past, given the Balinese a yearly balance in their favour of some two million dollars. This wealth has been rolling in for the last fifty years; but, the imports, during the five years that I have lived there, have increased 200%. It won't be long before these imports will wipe out the handsome balance in their favour. Then, the deficit will increase year by year, and the Balinese will find themselves in the same position as the other natives throughout the East. The cost of living will rise and the Balinese will have to work. In the United States we are so used to work that we can't conceive of life without it. We have placed work on a pedestal. It is our God, and anybody who does not do his share is immediately branded as a loafer, an idler. The Easterner—and particularly the Balinese—has found that life is infinitely pleasanter if work is reduced to its simplest expression and only indulged in when vitally necessary.

The slogan of the Balinese is, "Never do today what you can put off several days hence" but, when he has work to do, he does it in a most efficient and admirable manner. His fields are terraced and planted with infinite care, for he has discovered that the more care, the more yield. Once the work is finished, he retires to his home and thoroughly enjoys himself doing nothing at all, or working at his art. I have followed the same system with great success—at least when on the island. I have achieved the art of doing nothing in a magnificent degree and, somehow or other, am occupied all the time. Having leisure, my friend Spies and I started a scheme which would tend to slow down the invading forces from the West and keep the Balinese in their happy, contented ways for a few decades longer.

We saw with terror that the imports were increasing. We noticed that galvanized iron had crept in and that the natives, finding it convenient, used it for their roofing material instead of the thatch with which their houses were covered. We tried to show them that it was hideous. They did not understand, so we had to change our tactics. We struck upon something that we were able to drive home. We spoke about as follows:

"Pungawa (chief), you have recently invested in a magnificent galvanized-iron roof which covers your reception hall. That's very kind of you; you help our business. This metal happens to come from America, and in the name of our people we thank you for the kind donation. But, where did you get the money to buy this tin? You had to sell some cattle or pigs or copra or coffee. In five or ten years this galvanized iron will have rusted away and you will have to buy more to replace it. Once more you will sell one of your four commodities and buy some of this dreadful stuff. You are also, Pungawa, the owner of a motor car. It is an American one.

Again, thanks, but in four years time that car will be a rattle-trap, worth nothing. You will have bought a number of spare tires, many hundreds of gallons of gasoline at 45 cents the gallon, and lubricating oil, and your entire investment will be represented by some iron junk and air, plus a few pleasant memories. To buy this car you have sold a good many head of cattle and pigs or copra and coffee, and that money has left Bali forever— never to return.

"Now, my friend, five years ago what were your monthly expenses? Today how much are they? You admit that they have increased 50%. Why, you ask? Because of that very car you have bought; because of the galvanized-iron tin with which you have roofed your hall; the white goods that you are using as clothing, etc.

"Now tell me frankly, what advantage do you derive from owning that car? Time is not money here. You don't really need it. It's a luxury. Why that iron roof? Your thatch is cooler, lasts longer and costs less.

"Five years ago you used your own thatch, your ponies took you here and there without a cent's expense, and your women made your clothes. Your money stayed at home. If you keep this up, buying from the outside, in a very short time Bali, instead of being one of the richest lands in the world, will become impoverished and you, my dear Chief, will have to do like the people in Java—work for a living, nine or ten hours a day, for a mere pittance. You will become a nation of coolies. Do you understand that?"

Well, the Chief did understand. We spoke to other

chiefs, and to this argument added another. We said, "This is a feudal country. You people have been the chiefs and rajahs for generations. The white man has started schools. That coolie boy, if he is clever, may become the chief in twenty years from now. Democracy is slowly creeping in. Do you like that?" And, of course, they did not, for they know that a democratic government in Bali would be preposterous.

Then we spoke to the small folks and talked about the cost of living that had increased, and the good word spread so well that our organization is growing so strong that within a short time we shall be able to go to the Volksrad and demand certain changes.

We want to make of Bali a national or international park, with special laws to maintain it as such. We want a heavy import duty on automobiles, galvanized-iron roofing, white goods, etc.—in fact, on all goods not essential to the natives. We want to teach the villagers to make tile to replace the galvanized-iron roofing, batik, and block printing, and encourage all native home industries. Otherwise speaking, we want to keep the money at home!

We also want schools in which more attention will be given to native crafts and less to general instruction, which is of no earthly use to the natives. Some time ago I asked a young prince, who had spent six years in school and high school, which were the three greatest men of history. His answer was: Alexander, Karl the Great, and Reuter.

If he had been taught something about forestry or agriculture, he would have been better off.

We want a head tax on all tourists, and the money that

will come in from that source could be used as the Government may see fit—a sum which would easily off-set the loss of business of the importation houses.

These developments have come since Powell left the Island. The high officials before whom we have laid the plan all feel that we are on the right track, but they can do nothing to help us, since it would be against their business ideals—and more particularly against those of the business men at home! The officials are there to increase trade. They must show a higher percentage of business during their administration than during that of their predecessor. It is a question of dollars and cents, but they, of course, can see that an impoverished island will not yield as much as a rich one.

The very motto of "Bali for the Balinese" will help immensely. The Dutch do not move quickly, but they are extraordinarily steady. When they will have been "sold the idea," they will show their admirable good sense and assist in every way. For they have the welfare of the natives at heart and they are the greatest colonizing na-tion on earth.

We believe that a situation parallel to this one has never existed in the history of nations. It is a fascinating one. Whether the Balinese people will be able to keep their individuality, their customs, and their infinite hap-piness, depends upon themselves. All we can do is to point out the way.

ANDRÉ ROOSEVELT

March, 1930

NOTE

Bali, as usually pronounced by Europeans, rhymes with *jolly.* Of course, if you want to be a bit pedantic, you can accent the final syllable with a grunt, in the native Balinese fashion. But no white man can really grunt in the right way.

English-speaking persons always have trouble with the Dutch spellings of Malay and Balinese words; so I have Anglicized them, substituting *u* for *oe,* *j* for *dj,* and *y* for *j.* Thus Buleleng, Klung Kung, Ubud, Gianyar, and Jokjakarta appear on Dutch maps as Boeleleng, Kloeng Koeng, Oeboed, Gianjar, and Djokjakarta.

The vowel *u* is pronounced like *oo* in *soon,* and the other vowels are approximately as in French: *a* as in *father; i* as in *machine; o* as in *song;* and *e* as in *they.* Each vowel has full value in a separate syllable, and in general the final syllable is accented.

Lest any linguistic expert become hypercritical, let it be said that the Malay used in Bali is a bastard, pidgin tongue; and I have learned and written it largely by ear. In many cases I have spelled names phonetically. Ida Bagus Gidéh, for instance, would write his name *Id Bagoes Gde.*

H. P.

THE BRINK OF WISDOM

I

OUT in the Dutch East Indies, a week east of Singapore, a night east of Java, and just south of the equator, lies the little island of Bali.

I hardly know why I went to Bali. Perhaps it was to satisfy myself that other men had lied. For I was disillusioned. Months I had wandered in far places. I had seen the human scum of Soochow Creek. Among tall mountains I had watched the labours of the naked Igorote, crouched in his hut, and pitied the sooty poverty of his life. I had seen a dainty *vinta* skimming the sunset in the Sulu Sea. At its tiller was the grandson of a pirate; on its sail was the crest of a sewing machine company. I had seen a fierce Moro of Zamboanga, gallant in green and purple, stalking like a god to market with great fish of blue and gold. Then he climbed into his flivver, drove out past the cigarette billboard, to sleep in a sty. Off Borneo the sea at night glowed with a million phosphorescent opals, beyond a desert of corrugated iron roofs and filthy Chinamen. The air was sweet with frangipani and sick with the stench of market places. I had glimpsed the beauty of the earth's end. She was a slattern, soiled wench; and pursuit gave way to sad depression. I scarcely sought beauty any more; but now found Bali.

A clairvoyant, had one been with me the morning I stepped ashore from a small boat on the beach at Buleleng, would have sensed in my thoughts a certain sardonic satisfaction. I had suspected, but now I knew: travellers *were* liars.

Buleleng. Corrugated iron roofs. Chinamen in white pyjamas. Bombay traders, lost tribe of burlesque show comedians, with their shirt tails hanging out. Sweat and mosquitoes. Tin cans, through all the East the worst malfeasance of the Standard Oil Company. Dutchmen in peaked official caps, with high-buttoned choking collars.

The belles of Bali, where were they? I had seen men leer, and nudge, saying: "They don't wear any shirts." Men's voices had grown deeper and their eyes dim, as they spoke of chaste dryads in a tropic Arcady. Well, here they were, in their *sarongs* and tight-waisted, long-sleeved, sloppy *bajus*. I might as well have been in Sourabaya, port of Java, whence I had sailed last night. As I loitered in Arab Street, a slattern brown girl sidled by. She ogled me.

But somehow I found myself in a motor car, climbing a mountain slope. Up through step-like terraced fields, imprisoning the gleam of water, tinged green with spears of rice. Slender palm trees curved under great green nuts. The air grew cooler. Now we were among sheer cliffs and tangled jungle growth. The verdure blazed with poinsettia crimson. The air was like a well-honed edge of chilled steel. We burst forth into a place without foliage, a tattered, shaggy village perched upon a ridge. Before us was grey, black emptiness.

Far, far beneath shimmered a vast silver crescent, with tiny spots upon it; they were sailboats. Gunmetal clouds tumbled above it, and from its brink leaped a massive cone, which lowered down upon us from a summit wreathed in steam. Far below us also spread wide miles of jagged, broken blackness, spitefully steaming. André Roosevelt was beside me speaking:

"Two years ago I stood here on this ancient crater ridge, and from Mount Batur yonder fire was spouting to the very heavens. All its slope burst out in craters. Down below there, lay the village of Batur. The villagers knew their temple was sacred; ten years before, the lava had flowed to its walls and stopped. It would stop there now. Troops finally made them move. Then the torrent of lava came. A scientist drilled and sounded it the other day. It was a hundred feet deep, and down where the village had been the heat was still 900 degrees Centigrade. Up here is the new village of Batur."

Now we were rolling silently down beyond the mountain ridge, and suddenly the warming air was spangled with shimmering jets of sound. Before a temple men sat playing bells, and strange instruments like a xylophone. And wonder of wonders! it sounded as oriental music *should* sound—like muffled laughter of forgotten gods. Pleasant-faced brown men, women, and children had gathered around, and in an open space among them, between two towering parasols, a pair of tiny dancers trod a stately rhythm. Gold crowns flickered on their heads, and their gowns shone with pure gold also. Their hands wove fantastic patterns.

I stood spellbound, but then within me whispered the cynical, unconvinced thought of a tired traveller: "They certainly are doing a good job for the tourists. Let's move on."

As we descended, the air softened to a balmy warmth that kept the freshness of the mountains. Then appeared a solitary female figure, swinging toward us up the road. The sun shone russet on an earthen pot above her head, matched to the stripes of a bold *sarong* trailing easily from waist to feet. A scarf fell carelessly from a shoulder, and the bronze bowls of maiden breasts projected angular, living shadows. She walked majestically, with slowly swinging arms, with never a glance for staring eyes that now rolled past her. And all at once she was but a part of a vast spreading wonderland, embodied dreams of pastoral poets.

South Bali lay before us, a teeming, pregnant woman, and in her eyes burned afterglow of fallen empires.

Eighty-two years had passed since men-of-war came to the north coast, bristling with guns, and brave columns came down to meet them, with the lance and *kris*, to die upon the smoky strand at Buleleng.

Eighty-two years ago, but not until the turn of the century did someone get worried about morals. Some say it was the Dutch Resident, concerned about the minds of adolescent sons. Some say the Commandant was anxious about the manners of his troops. At any rate, an order went out. For purity's sake, the women of North Bali submissively covered their breasts with the *baju,* hitherto the badge of prostitutes.

While North Bali was donning the shirt, symbol of subjection, a war fleet came to the South, near Den Pasar. White men landed beside a coral temple. They stood in battle array before a native stronghold.

A strange spectacle came before them. Out from the town stalked men in crimson, bright with gold embroidery. They carried lances, and gold *kris* handles glistened between their shoulder blades. Borne aloft by coolies on a gilded chair, rode a Raja, gallant in royal robes. A priest led the column, and strangest of all, it included many women. Glittering with spears and bared *krises,* it advanced toward the invading army. Defensively from the invaders burst a roaring, fiery volley, raking the line with death. The Balinese halted.

But now the priest's *kris* was flashing. Gold damascene shone from its slender blade, wavy as a serpent. It plunged into the bare, round breast of a woman. Another woman, and another, fell before its lightning strokes. Now all the men were striking down their women. They slew their wives and then they slew themselves.

Thousands had died in old campaigns, valiantly, madly, before the blaze of fire-arms. It was useless. The end was come. The Dutch stood facing a proud army, dead but unconquered.

In the fields of South Bali the rice growers went on with their labours, and lived their lives just as always before. They always had submitted to authority. Now there was a new authority.

The mountains reach up to the clouds and bring down floods of rain. The mountains, trembling and groaning in the night, spew up mud and ashes. And the waters wash down Nature's fertilizer, trickling from terrace to terrace, toward the sea.

We were rolling down through the terraces, into the teeming, pregnant South. Everywhere were hills of coffee, coco groves, vast terraced vistas of rice. Dutch telephone wires straggled from planted poles that had sprouted, spread into flourishing trees. Everywhere was life. Ducks were swimming among the rice. About the villages were many swine, strange creatures weighted to collapse with their great loads of pork; their backbones sagged as if broken. Everywhere were cattle, with sweet-toned wooden bells—lithe agile animals that leaped upon the banks with all the grace of antelopes.

Everywhere was humanity. Roads streaming with girls. Heads proudly bearing burdens. Ankles, elbows, balanced curve of breast and armpit. Soft eyes, hair negligently hanging. Deep-shadowed backs of laden coolies. Breasts of aged, brittle women, and of fragile, unsexed children. Everywhere life, swarming, seething. Life, surging in the market place beneath a massive banyan. Youths lolling by the roadside, fighting cocks brilliantly preening. High-held heads with baskets on them, chins curving into necks, and the tapering curve of fingers. Everywhere motion.

Bare feet, padding softly. Arms, in flickering lights and shadows. Lips, suddenly smiling. Bronze figures gently moulded, statuary bending in a smooth, unhurried dance of life.

The road ran on and on, a wide avenue between stone walls. Everywhere temples lifted their stone gates, carved as feathery as the banyan trees above them. The villages were miles of walls, thatched against the rain, with hundreds of prim pillared porticos, and groups of damsels sitting by them. Beyond those parapets were homes. What sort of people lived there? What manner of life did they lead behind their sheltering barriers?

If, with Roosevelt and me, you had travelled this route one morning in April, 1928, you would have gone from Buleleng 129 kilometres to Den Pasar, the capital of the South, then on a mile to the town of Sumerta, which is a group of nine *banjars,* villages. You would have stopped before a gate in the *banjar* of Bengkel, and there, amidst a gay welcoming group, you would have beheld the smiling visage of our friend Kumis.

II

PUT a shirt and trousers on Nioman Kuko, and you might take him for a sunburned Sicilian of noble class. A high brow sloped back from his straight nose to a shock of silky black hair, topped by a head-cloth of bold batik design. His batik *kain,* which hung from waist to shins, had blue vines and crimson butterflies. Smooth muscles rippled on his bare bronze chest, and from his lip curled a pair of black moustaches. Because of this distinction, his neighbours called him Kumis—"Moustache."

Andre Roosevelt and I were always welcome in Kumis's house, for an afternoon call or to spend as many nights as we wished. He was always ready to sit around smiling and talk to us, smoking our cigarettes. Or sometimes he would just sit silently, sociably, fingering his smooth chin. Occasionally he would find a hair, and pull it out, with a pair of iron tweezers.

I never saw him do any work. He had many rice fields, and at planting time and harvest he would be busy out there, but now there was nothing much to do. Once I did see him, between the shafts of a big bullock cart, dragging it down the road. He waved gaily at me and cried, "Just like a coolie!" He was always in evidence when anything was going on around the village of Bengkel, but he never seemed to have anything to do with the proceedings. I had known him six weeks, and Roosevelt had known him two years, before we learned that he was by common consent head man of the village.

As likely as not, as Kumis stood by the gate to greet us, he would have his bronze fighting cock in the hollow of his arm, its fine tail curling over his elbow. A staunch and wiry bird, this rooster, with a high flaring comb. He was very fond of it.

Kumis had eight fighting cocks. He kept them in bell-shaped basket cages, each separately. Sometimes he would take them out, one at a time, and fondle them. He would stroke the bright long feathers of their ruffs, and let them walk about and exercise. Sometimes he would pair them, and let them spar awhile, ruffs aflare, feathers flying. They all had won their fights, he said, but the bronze one was the best. The Dutch prohibited cockfighting, except on special holidays, and immediately it ceased, so far as anyone could see. But in the villages, where there are many secluded enclosures, the cocks remained as ubiquitous as hip flasks in a Broadway cabaret. Every morning Kumis would range his roosters' cages in a row outside his gate. He did that, he said, so they might see people going along the road and be amused. I think he pitied their rigorous, Spartan lives, their monkish celibacy.

Kumis himself had two wives. Neither Madé Renkog nor Réntung was as handsome as he. They were getting

along into their thirties, and were fading. Réntung, I am
sure, had been pretty. Madé probably never had been.
Consider her features one by one and you would conclude
that she was very homely. Her face was long and narrow,
with large, heavy lips; and when she smiled she showed
ill-shapen teeth. She was a bit thin and angular. But there
were her eyes, and the soft hair that fell negligently about
her brow. Eyes everywhere were beautiful in Bali, but there
was about Madé's a brown depth whose immensity seemed
to encompass all the soothing peace of this gentle island.
When you looked at her eyes it was hard to see anything
else about her. She would sit there, smoking a cigarette;
and I, caught in the spell of her repose, would stare at her,
wondering how she could be so ugly, and yet so beautiful.
Perhaps it was her gentleness that transfigured her. I
liked her best when little Babad would come and stand
close beside her. He liked to be there. Without touching
one another, they seemed to be caressing. She had no child,
but she was born to be a mother.

Babad was ten, and always wore a brighter *kain* than
anybody else. Once I took to Kumis a new *sarong,* of very
fine *batik* with blue and yellow peacocks. The next day
Babad came running home from school with it wrapped
around him, tucked up about the waist. Under his arm
were bundled his slate and his little white shirt. The Dutch
made him put on a shirt when he went to study Malay and
arithmetic, but he did not like to wear it. He was a quiet
little fellow, with a wide white grin. His hair was clipped
short, except for a long lock in front. Sometimes Madé

would twist this, tie it in a knot to keep it from his eyes. "He would get sick if it was cut," she said.

It was too bad, I suggested one day, that Babad had no brothers and sisters.

"He has many brothers and sisters," said Madé. "They live in a house down the road. Here, we had no child; so we took him."

Kumis's home, like a hundred thousand others in Bali, was a walled enclosure. It was about forty yards square, and within it were seventeen small buildings. At the rear it opened into the family temple, some twenty yards square, a simple place, without much ornament, but good enough for family worship. Most of the little houses were grouped around a square in one quarter of the big compound. Here lived Kumis's sister and her family. There were fifteen altogether in the household. Kumis had the rest of the big lot for his own quarters, cook shed, and granary.

Kumis's personal dwelling was at the rear of the compound, a little white plaster house with door and shutters that closed tightly. It had a front porch where hung two withe cages, with cooing doves. There were painted decorations inside the cages, where the doves might see them. This house had a red tile roof, but all the others were topped with thick smooth grass. Some of the other houses had bamboo walls, but many of them were open to the breezes. The floors of all of them were built up with bricks, a yard above the ground. This assured a degree of privacy from the pigs and chickens that roamed about the clean sand of the yard.

In the exact centre of the compound stood the family living room, the *balé balé,* and it was this that Roosevelt and I occupied in the many days we visited at Bengkel. This *balé* had a thick thatch, and its elevated floor was shielded from sun and rain by broad grass eaves. In it were platforms, like beds, of springy split bamboo. It measured about fifteen by twenty feet. One end was closed with a wall of brick, and mattings hung on one side. The other sides were open. Kumis found two mattresses for the bamboo springs, and the village priest sent us a table and some chairs. I must confess we were not sorry the influence of the West had penetrated to this extent.

Nor did we regret that Réné, who lived down the road, had been servant to a Dutchman, for she came to cook for us, squatting at a little charcoal brazier at the corner of the *balé.* She was rather ugly, with slender, boyish form, pop eyes, and kinky hair, as though she had Papuan blood from farther east. She was the only woman in Bengkel who affected a shirt, and she devilled us until we got her a pair of leather sandals. But she could make a good curry, and broil a chicken well. She thought we were very extravagant when we bought chickens, at two for a quarter.

The gate of Kumis's home had no door to close it. A low bamboo fence barred small animals, but otherwise the entrance was free to all. The entire village flowed in and out, and always at our *balé* we had a sociable group of callers. They were there when we went to bed at night, and when we got up at morning, to admire the automatic stropping of my razor. It may seem we had scant privacy, and to tell the truth there is very little of this to be had in a country so teeming as this. But one soon learned to turn one's back and ignore the world. One soon learned that modesty was a quality of the mind, and that no person was more modest than these half-clad villagers.

I was lying on my back one day, looking up inside the roof of this simple dwelling of ours. Matched bamboo poles converged like rays from the spreading eaves, upholding yellow grass, all precisely bunched and tied with split bamboo. The poles came together at a heavy ridge beam of hard wood, upheld by a wooden pillar, expensive wood that had to be imported from Java. These were intricately carved, with a fanciful flower design. It all made a charming picture.

"Who did that carving?" I asked Réné.

It was Ktot, that young man who was here last night with the artfully fashioned *naga's* head of wood, she said.

"And how much did Kumis pay him?"

"Pay him? Nothing. Kumis bought the wood, and Ktot carved it for Kumis."

Here we slept through glamorous nights, and except for places in the mountains it was the only spot in the tropics where I found a mosquito net was quite unnecessary.

While I was staying in Kumis's house it usually took me a long time to get dressed in the morning. There was so much to see, going on around the dooryard. I would while away hours during the day, just sitting there watching. If you can imagine a three-ring circus, all pitched in a soothing key, you will have an idea of this home, of all the life in Bali. It was not languid, but there was never any hurry. Things seemed to get done by chance, and there was never any bossing.

Madé would be cooking rice in the open-sided kitchen, with its brick fireplaces. Réntung would come from the well, with a pot of water on her head. Someone would be sweeping up the yard with a bundle of twigs. In the shady place under the peaked roof of the granary, beautiful Renang would be sitting all day weaving.

Kumis had a little niece named Nioman Renkog, Renang's sister, who was ten years old. There was a flat stone by the corner of our *balé,* where little Renkog would make medicine. She would come with ginger root, and all manner of herbs and dried flowers, and a strange root that made everything bright saffron. She would put them on the stone, wet them, and pound them with another rock, until she had a well-mixed handful of yellow paste. This was very good, especially for babies, to rub upon the brow and breast, to keep the ills away.

Renkog would help Madé pound rice. She had a little wooden mortar and pestle, just like Madé's big ones. In a big bowl hollowed in a stump of wood Madé would put her grain. Her long pounding pole was heavy, and shod with iron. She seized it with her right hand, balancing it vertically, and flung it downward into the pile of rice. Her left hand caught it on the rebound, lifted it, and flung it down again. So it rose and fell, in a tomtom rhythm. Sometimes Réntung would pound also, her pole beating alternately with Madé's in thudding steady time. Chickens darted in to pick up flying grain. When the pounding was done, the rice was tossed in a great wide basket tray, and chaff flew in the air like powder.

Little Renkog was learning to walk straight, majestically, with loads upon her head. She was learning to pound rice with balanced swinging of the arms. When she grew older, how comely her body would be. As comely as Renang's.

Meal-time was just like Sunday morning breakfast in any well-regulated household. Everybody ate when he felt like it. The food was rice, bananas, and other fruits, coconut grated and seasoned with red pepper, sometimes a bit of chicken or pig, well spiced. It was eaten with the fingers, from squares of the thick green banana leaf. When these had been used, they were tossed out for the pigs to eat. There was no dish washing to bother with.

Madé was always cooking some kind of green leaves into soup for the pigs. She would feed them at noon, pouring the food into great stone bowls around the kitchen. There was a bowl for the big pigs, and another for the little ones. They were clean pigs, not kept in any filthy sty. They were curiously built, with sagging backs that almost dragged their stomachs on the ground. They ran about the clean yard, and if there were any edible débris there, they would eat it.

Three cows were stabled in the corner of the yard, amiable, graceful beasts. In the morning they were taken outside the walls, into the grassy aisles among the palm trees. In the afternoon Babad would take them down to the stream which ran past Bengkel, throw water over them, bathe them well. These cows, like most cows in the tropics, gave no milk. They were sacred, and it was forbidden for the Hindu to eat them.

But now and then a trader would arrive and buy a cow and some pigs from Kumis, and some from his neighbours, and ships would go out from the island laden with living things. They would come back with great chests of silver *ringits,* like American dollars. Kumis would pile his *ringits* in neat stacks, according to the dates engraved upon them, and bury them in the mud beneath his house, with many others, just as millions of other dollars were buried beneath houses all over Bali. They would come forth when there was need for them.

Kumis's gate opened into a wide straight street that had been grassy until the Dutch put gravel there. It was lined with walls that had thatch on top to protect them from the rain, and every little way there was a prim portico like Kumis's.

Just beyond Kumis's the brook meandered, down through its spraying clumps of tall bamboos; and beyond it was the village of Kepisah for a hundred yards.

Then the road came out into an open space, and far, far away amidst a feathery crown of white arose the cone of the Great Mountain. The road ran on amidst vast shelving fields of rice. The terraces themselves were tiny patches, each bound within its curving wall of earth, and dropping shining cataracts each to another. It was just after planting time when I first was there, and the shining terraces were faintly tinged with green. Two days later the blades of rice were an inch above the water.

Kumis had many *sawahs* out there, I don't know how many. When planting time came he would pull his *kain* up, so it was just a loin cloth. He would get down the wooden plough that hung in his granary. He would take his cattle to the terraces and hitch them to the plough. It was something like a sledge, and he would stand on it to force it down into the watery ooze of the paddy patch, then drive his cattle round and round, stirring up the fertile mud. He would get some of his neighbours to help him, too, for a share of the crop. When the terraces were ready Madé and Réntung would come out to help him plant the seedlings.

There wasn't much more work to do for three months. Then when the fields were yellow and dry the whole village would go out for the harvest. Almost every man had his own land, and those who had little helped the others. Everybody shared in the crop. Each got a third of what he gathered. There was plenty of food for everyone in the village. What there was to spare was shipped away.

Such little things as Kumis's household did not produce for itself could be bought at Kepisah. Weekly the women of the neighbourhood gathered there in the market place, sat shaded under mats, happily exchanging vegetables and gossip. For greater needs there was the great market at Den Pasar every week, to which came all the women of the country around. Kumis's wives had a good time on market days.

It was strange to watch the market women walking

home, with their wide square empty baskets on their heads. One corner of the basket would rest on top the woman's head, and the rest would stretch out over the woman's shoulder. It seemed to defy all the laws of balance. But that corner atop the head was weighted with the day's takings. Piled there were bronze Chinese cash of the Manchu dynasty, worn thin with many years of trading. It took 1200 of these to make a *ringit*.

Kumis and his family were rich. They had no shoes and no shirts. But they had everything they wanted, with some left over.

They had not read the *Saturday Evening Post* and learned to want a washing machine, vacuum cleaner, radio, motor car, asbestos shingles. They had not read the *Saturday Evening Post* and learned to expect or want life to be full of thrilling impossible adventures.

Since I left Bengkel I have heard a lot about Prosperity. Poverty is being abolished. The man who has a Ford and wants a Cadillac is not poor.

Our standard of living has been raised. Some of my friends have electric refrigerators, motor cars, and many other things, and their wives work in offices to pay for them. Their wives like nice things. The abolition of poverty has raised the standard of living so high among some of my friends that they can't afford wives, not even one.

III

I HAD come to Bali to spend three days.

I had found a tiny spot that is different from any other place in the world, inhabited by humans who might, for any similarity with our northwestern life, be men from Mars.

There was magnificence in these latitudes a millennium ago.

The mysterious Hindu Khmers came down from the body of Asia into peninsular Cambodia, built a city greater than the Athens of Pericles, and vanished, leaving the glory of Angkor to rot in the jungle.

The Hindus came to Java also.

"In this country they have made the city walls of piled-up bricks, the wall has double gates and watch-towers," wrote a Chinese voyager who went to Java fourteen centuries ago. "When the King goes out he rides on an elephant. He is surrounded with flags of feathers, banners, and drums, and is covered with a white canopy. The people say their country was established more than four hundred years ago."

For a thousand years more the island of Java flourished in the great empire of Majapahit. The walls of its city were thirty feet high, its standing army had thirty thousand men, and in far Sumatra a descendant of Alexander the Great swore fealty to the ruler of Majapahit. Massive, beautiful temples were reared in Java.

The Hindus came to Bali also. They left relics of magnificence. You can go down a causeway cut through solid rock, into a deep canyon. And there, against cliffs honeycombed with a rock monastery, stand ten great cenotaphs, in memory of forgotten kings. But there are few ruins. It is the living people that remain, and they still build temples.

The city of Majapahit in Java has completely vanished. Five hundred years ago, Mohammedanism, filtering along the routes of trade, smothered the culture of the islands. The Arab and the white man who followed him brought an era of colour, adventure, and romance, a fiery fighting era; but the sword of Islam has grown dull in the seas of the East. The barque and the fighting *prahu* have vanished from the seas of Almayer and the Outcast, given way to the unromantic, if excellent, steamers of the Blue Funnel and the K. P. M. Joseph Conrad sang their song, and now he too is dead. Up north of Borneo the Sultan of Sulu, against whose forebears the Spaniards walled Manila, is a fat old man in a tawdry hut. He shook my hand with a paw grown flabby in the counting of British and American gold, and returned drowsily to his betel box, his electric flashlight, and his blackboard lesson of A B C's. His wives are childless. Strip from the sea rover his fierce audacity and you have a sad spectacle. It is not pleasant to look upon the corpse of romance.

But in two places a glamorous past still lives. At the royal courts of Jokjakarta and Surakarta in Java, the Hindu culture remains, a cloistered flower of aristocracy, scarcely touched by Islam. And in Bali the farmers make music, the field hands fashion stone, and coolie girls dance in cloth of gold.

The survival of Bali verges upon this miraculous, but has its explanations. The island has no harbour, and its iron-bound coast, which holds off tourist ships, for centuries held off trade and the changes trade brings. Bali's brave men also had sharp steel, which for centuries held off all invaders. It was not easy for the Dutch to conquer them. Expedition after expedition was repulsed. Thousands upon thousands gave their lives in defence, but their blades could not withstand the white man's firearms.

Once captured, Bali became a negligible jot in the vast empire the Dutch were taking in the island-scattered seas between Asia and Australia, an empire fifty-eight times the area of their native Holland and extending (from Sumatra to New Guinea) as far as Lindbergh's famous flight. At the south central border of this empire, well south of Borneo, at the north edge of the Indian Ocean, 115 degrees east of Greenwich and 8 degrees south of the equator, lies Bali, left pretty much to its own devices, governing itself with the advice of a few Dutch officials. The Dutch, who are probably the world's greatest colonizers, are busy exploiting Java and the other large islands. They haven't yet got around to Bali, and it may be a long while before they do. The Resident is an enlightened man, as is the Governor General. So far the Dutch have contented themselves with a little bit of revenue, and have left the lands in the possession of the people.

Thus Bali goes its own peculiar way, a land that has no struggle for existence. The only struggle it ever knew was war, and that now is taken from it. It never had an economic struggle. Here hungry nature, well fed, smiles. This little island, slightly larger than the State of Delaware and largely mountainous, supports on its fertile lower slopes, in abundance, with very little labour, a million people. Here are food, health, laughter for all, and leisure without benefit of mass production and the five-day week. Its people have everything they want, and no advertisements to make them want more. They have leisure that is filled without the aid of instalment plan phonographs and flivvers, filled out of the rich imagination of an energetic people, with art, music, dancing and great religious festivals.

The western atrocities that move in the van of commercial invasion have flowed into the lives of many peoples as into a vacuum, but southern Bali is almost untouched. Its people are satisfied as they are, they are remote from the flow of commerce, and their life is already full of their own civilization.

Yes, civilization. You don't think civilization is only a matter of toothbrushes, typewriters, and stock dividends, do you? There are other civilizations. You don't share, do you? the viewpoint of that Charleston-dancing young man from age-old Siam, slightly ridiculous in his baggy trousers, who told me with an English accent: "Oh, yes, my country is getting quite civilized."

Our own civilization has its vehement critics, with whom I cannot go all the way. "Getting and spending we lay waste our powers," cried Wordsworth. The men of Concord are enslaved by possession of their farms, cried Thoreau. Man is being enslaved by the machine, cry the critics of today. They have it tangled. If the man of our western world is enslaved, it is by heredity. For what are we bred, if not for possession, for getting and spending? In our unfriendly, cold-wintered world, what blood would survive that was not acquisitive? Millenniums of breeding have flowered in Babbitt. Babbittry is the one touch of nature that makes our whole race kin. Do you object? Do you speak of artists, musicians, writers? Show me one who is not out to get his, or to "get somewhere" and I will show you a mistake of nature.

Babbitt will inherit the earth. Babbitt, who already has started on the north of the island, will ruin Bali. And all I know to say of that is Kismet. But meanwhile the life of Bali is an anachronism in the era of Henry Ford. People travel many miles to visit the ruins of the ancient Hindu empires of the East, treasures long forgotten and corrupted in the jungle. But here in Bali that culture still lives, the vigorous, exquisite treasure of an entire people.

I went to Bali, as I have said, to spend three days. As I stood reluctantly on the beach at Buleleng, waiting to be rowed off to my departing steamer, Roosevelt said:

"Stay around a while. I came for three days and have

been here three years. You won't mind sleeping with graven images, will you?"

So I put down my luggage in a room cluttered with stone demons and wild paintings. My boat sailed and I caught a duck in the courtyard at Roosevelt's house in Buleleng. We had him for dinner, with black rice and American-style coffee that would warm the heart of any jaded traveller from Java. We talked about Bali and beauty and men who worship strange gods, and about New York and machinery and men who worship Things.

We were happy to be in Bali.

IV

IT is my first night in the house of Kumis at Bengkel. Roosevelt and I have just finished eating a good curry, and are feeling very homey and comfortable here in our *balé balé*. Our murky little oil lamp casts strange shadows about Kumis and Madé sitting there.

Out of the darkness comes a girl, and the dim light gleams bright on her eyes and teeth. It is Renang, who has brought fruit for us. I catch a glimpse of fluttering hands. She vanishes.

The fruit is in a Chinese dish. Two hundred and fifty years have worn away its glaze. It is a Ming plate.

The fruit is mangosteens. I break a soft maroon husk, and within is a little white globe, segmented like an orange. Tropic fruits are tropic fruits, but they include the mangosteen. I close my eyes, treasuring its honeyed savour.

Now out of the silence drift faint sounds of drum and cymbal, and a reverberation.

"*Kumis! Di mana gong?*"

"*Tempat gong ada Kepisah, tuan.*"

Kumis leads me out and down the village street.

The world is wet with moonlight and the honeyed drip-

ping of the stars, and vacant aisles among the palm trees
yearn for nonexistent lovers. The warm scented air is
throbbing. It is singing the Song of Songs. "I am black but
comely," sings the night. The night cries out for love. But
now upon it bursts the boom and tinkle of brass, like a cold
astringent douche. This intricacy of tone which pours into
the tropic night is of an ascetic, cerebral quality. It falls
pleasantly, but at first incomprehensibly, upon my occidental ear. But as I listen intently I know that this is a
highly developed counterpoint, based on the simplest of
melodies, simple things exquisitely interwoven, like the
pattern of an oriental rug. From the simplicity of x and y
are built all the intricacies of quadratic equations. And no
matter how much you may loathe it, algebra is beautiful.

The language of Bali has no word for love, nor has this
music any theme of passion. In its brazen percussion I
search in vain for any hint of the loftiness of Beethoven or
the sentimentality of Mendelssohn, for any emotional content at all. It is played in the midst of the most ravishing
beauty, by men in close affinity with nature, but in its measures is no hint of that—no rippling brooks or winds in
treetops. There is no literary quality in this music. It is
tonal patterns.

Kumis brings me to a place where a tower stands sil-
houetted with the palms. In its belfry hang three logs,
with slits in them that let the starlight through. Beat a
tattoo, they will sing their own distinctive tune. One time
they called Kepisah's men to war, but now are silent. This
place is the *balé banjar,* the men's club, the civic heart of
Kepisah.

From a house with woven bamboo walls the music comes. We climb its steps and enter. Dim lamp-light reveals two dozen bare-chested men and boys, sitting on mats, ranged in a square. Before them are wooden racks, hung with bells and bars of bronze, and their hands fly with little hammers. Opposite me, on high frames, hang the great bronze gong and its smaller brothers. At a rack of thirteen pudding-shaped bells stand four men with fluttering sticks. One man sits with a long drum across his knees, his fingers caressing its head, and in the hollow of his arm leans a naked baby. Two boys about ten years old have each five bars of bronze, and their little hammers flicker, their left thumbs following close behind to silence staccato tones. They never had music lessons. They learned like the drummer's baby. An old man, his bronze suspended above bamboo sounding chambers, somnolently strikes sustained bass tones. The musicians sit with vacant, inattentive faces. The men of Kepisah play as though entranced, spinning their kaleidoscopic web of sound.

It requires concentration to listen. This music is the first I have heard which sounds as oriental music should sound. But to ears attuned to the broad contrasts of woodwinds, strings, and brass it is at first a bit confusing. It requires an accustomed ear to catch the varied voices of this percussive bronze. And there is nothing of song about this music, just short themes mingled and repeated with infinite variations.

The gamelan of Bali is a close cousin to the orchestra of Java. I am told that though the temperamental char-

acter of Balinese music carries it to peaks above the Javanese, the Javanese gamelan generally plays on a more even, fastidious plane than that of Bali. I don't know about this and don't care much. But this I know. During ten casual days in Java I heard no sound of native music save the atrocious ringing of bamboo bells by small boys who hoped to pick up tourist pennies. During my first day in Bali, motoring along a narrow strip of road, I saw and heard three orchestras. The life of Bali is a continual series of communal events, and at all of these is music. At times it has seemed to me that I might stand at almost any spot in South Bali, at almost any hour, and hear the distant sound of music.

When a Raja of Ubud was to be cremated, it was asked that as many orchestras as possible from that small district appear at the obsequies. One hundred and twenty-six gamelans appeared. Every community has its orchestra. The musicians are not professionals. If the orchestra be paid, the players themselves get not one cent. They learned music because they loved it. They are farmers. These men of Kepisah make their living by ploughing rice fields, tending cattle, herding ducks; and they take their recreation in the assiduous practice of a very high form of the highest of the arts.

The *gamelan gong* of Kepisah is playing. There is no conductor. Now and then there is a second's pause. The treble *gangsas* strike up a new theme, seeming to follow that youngster with the shirt. Now their hammers fly in independent unison, and the ensemble joins in with all its

contrapuntal multiplicity. A half hour passes. On and on
they play.

*How to describe the music of the gamelan, to ears which
have heard no semblance of it? I may say it is virtually
all in a minor key, virtually all percussion; but then what?
We who have written about music have adopted a jargon
of stereotypes to approximate its idiom. We speak of
music as pensive, gay, as filled with morbid despair or
spiritual exaltation. And whether it is that way or not, we
have contrived to give some semblance of its nature. What
does it matter if in Franck's symphony Van Vechten hears
a choir of angels mounting unto heaven, where I hear the
limpid murmurings of a passionless love? Each of us has
projected a pale shadow of its reality. But we know all
the time that it really means nothing when we speak of the
love songs in a Tschaikovsky symphony. We may con-
descend toward program music and affect to despise it,
and render homage to the creators of fugal patterns, but
how we thank the muses when Prokofieff exorcises devils
or someone else writes a tonal biography of the 10,000,-
000th Ford! We may disapprove, but at least we have
something to write about. The Balinese musician gives no
such aid. He plays pure music, which by its very existence
is romantic, but has in it not one jot of romanticism.*

*There are in Balinese music rare points of tangency
with occidental tonalities. It does not do much good to
say that the gamelan is built of instruments like a xylophone; for the scale is different, as is the tonal quality.
But in that group of instruments which has bamboo sound-*

ing chambers to mellow its brazen percussion, you might find a kinship of tone with the gipsy cymbalom, from which Liszt gleaned his rhapsodies. And the ear clutches with delight at the familiar pastoral notes of that great bamboo pipe which is part of the gamelan *for the dance drama. Close the eyes and from that gentle fluting it seems that at any moment may burst the voice of an Isolde, singing the impassioned lyrics of the* Liebestod. *But Bali is far from fat sopranos. Instead now comes the childish treble of a tiny dancer. A short, plaintive, minor tune, half spoken. No, it is not a whining. No, it is not off pitch. It is only that these ears are attuned to a different, less symmetrical, differentiation of tones than our own. Listen intently and you will perceive that in this strange sequence of sound is a precision and artistry quite fit to be considered a* bel canto.

But there is no singing or fluting now. This is the gamelan gong *of Kepisah playing.*

On and on play the men of Kepisah. They have no written music. They learned by ear. Their music never has been written down. This is but one of many long compositions. What manner of minds can hold it all? No one ever forgets or makes a mistake. But yes. Now the drummer seems dissatisfied. The music stops. The phrase is repeated. No, he cannot get it right. Three times they try it over. The group grins good-naturedly. A *reyong* player says a bantering word, and the drummer joins the laugh. The *trompong* player puts down his sticks and takes the drum. His fingers blur on the heads like gnat's wings.

They break into strange, incredible syncopations; they flutter in rhythms as blithe as Mozart's. That's the way it is done! The drummer smiles appreciatively. He remembers now. The music goes on.

The *gamelan* weaves its bewildering web of sound. A standing man strikes that great cupped bronze disc, which is the gong. It booms beneath its smaller brother.

How to describe the timbre of the gong! Take the smooth, round tone of the temple bell that sings through the sacred grove at Nikko; pucker it a bit with clang. Do you know the taste of ancient brandy mingled with claret? The sound of this great hanging bowl of bronze is something like that. It comes to the consciousness suddenly, at the uttermost depth of sensibility, with gentle stunning impact; it spreads with slow warmth through the veins; it tingles in the finger tips. It reverberates through a measure.

Upon this foundation is built the Balinese architecture of sound. A strange structure to our ears in this: that its intervals of tone are slightly different from ours, in a way that cannot be accurately written in our notations, that what might be called its octaves are never exact octaves but what we would consider a trifle off pitch. Now do not shudder, you who wince when Galli Curci sings a phrase off key. It's not like that at all. You will not feel that pleasant consciousness of critical superiority that accompanies your groan when Gigli muffs a high note. A better comparison would be a thirteenth century cathedral. Its lines are not straight or parallel, and the pillars of its nave and cloisters are set in irregular curves. Its very lack of apparent precision is the source of much of its charm, the quality of a free hand drawing. That tremendous chord of the gamelan gong, when its semblance is played on a piano, is a complex but commonplace dissonance. Struck by the gamelan, it is magnificent.

Rest period. The Kepisah players pass me a tray heaped with scrap tobacco, *siri* leaves, and cigarettes. Genially we smoke together. They tell me the names of their instruments. The *trompong* player lets me ring his ten pudding-shaped alto bells. They all laugh merrily at me, and Kumis claps my shoulder in applause. Now they play again. I listen for a half hour, but I have been abroad since daybreak. I am tired and must go to bed.

A gentle rain is falling as Kumis and I walk home. The night is murky, but it seems that I have menthol in my brain. How cool, clear and simple is this emotion-foggy world, when one has heard the singing of the spheres.

As I sink into sleep, I hear the booming gong. The men of Kepisah arose at daybreak too, but they will play on far into the night.

It is plain to Kumis that I love music. In the morning when I wake he beckons me into his house. There, polished and bevelled, glitter dozens of brand new bars of bronze.

"The music smiths of Klung Kung made them," says Kumis. "They are Kepisah's."

Kumis leads me back to the *balé banjar* of Kepisah. It

is a busy place these days, for the musicians are making instruments.

The *gong* of Kepisah was organized only a few months ago. Kedaton and Bengkel, next door, both had fine orchestras and dancing clubs, so surely it was time that Kepisah expanded its little *gamelan* into a symphony of sorts.

The men of Kepisah have practised much, and already they have a considerable repertoire. Of course, it was not a matter of starting from the beginning. They had several accomplished players already. All of them had been hearing this music from infancy, and there was the place of the *legong* just down the road, where in childhood they always had been welcome to stop for a while, play with the instruments, and learn the elements of music.

Certainly, however, there was much to learn. All the details of these delicate sound embroideries must be learned by ear. The instruments were simple, but there were many tricks in playing them. The fingers patting the ends of that long drum could bring from it four distinct tones, and the four-four time that is always used could be splintered into a multiplicity of rhythms. There were many tones to be made by the four men who played the thirteen inverted bowls of the *reyong*. They might strike the knobs with the padded ends of their sticks, or play upon the edges, or tickle the bowls with the bare handles. The neighbourly men of Belaloan sent one of their number to teach Kepisah all these things.

The instruments with which they started were just crude things picked up here and there, not at all fit to represent the art of Kepisah, and fine new ones must be made, with the bars that are stored in Kumis's house. Kumis is helping to pay for them, though it is not his village.

A half dozen men are working today, and Kumis smiles with delight as I crouch to watch them closely. They are making the *jegog*, the *joblag*, and the *jalung*, which strike three mellow octaves of baritone and tenor. On a solid base of teakwood, imported from Java, they erect two solid pillars, curving into horns. Between these are ranged five vertical tubes of bamboo, varied in size. All are beautifully joined and held with pegs of wood.

Old Mudari has nearly completed one instrument, and on cords of rawhide has strung the bronzes over the bamboo sounding chambers. I strike one with the padded hammer and it sings a sustained note, deep and mellow, with just a tinge of clang. A touch from my finger and it is silent. I play the scale. It is of five notes, which might be called mi, fa, sol, si, do or E F G B C, but each note is just a trifle sharp or flat from its closest counterpart in our occidental scale. As I play, Mudari listens. That second note, he observes, is just a little off pitch. He must remove the bamboo and change it.

Old Mudari knocks off work to talk to us a bit. His teeth are pretty bad. He puts some *siri* and betelnut into a joint of bamboo and tamps it with a stick. It is masticated well. With a little paddle he spreads it on his tongue, with a grunt of satisfaction.

Mudari has a request to make. When we return in our

car to Buleleng, may he ride with us and take the gongs? The gongs, from which the *gamelan gong* takes its name, came from Java. The music smiths of Klung Kung can make the smaller gongs, the *kempur* and the *kempli;* but for the two great ones, a yard or more in diameter, Kepisah had to go abroad. Three pair were sent from Java, and the musicians chose these, which went best with their other instruments. They cost $300. But there is still an adjustment to be made, says Mudari, and he is taking the gongs to the makers in Java. Just a little chipping and filing and the combination of the *gamelan* will be perfected.

Before long, after the gongs return, the *gamelan* will be complete. It will have three octaves of *gangsas,* tinkling treble without sounding chambers. It will have drums and little cymbals, and a *trompong* with ten alto bells. All will be matched with each other, in a complete instrument, unlike any other in Bali. But it will not be finished.

For the *gamelan* of Kepisah must be beautiful, worthy of the wood carvers of Kepisah. It will be cut with flowers, with leaves and petals and vines all intertwined. Its carvers will have no models, will scarcely sketch a plan. No two parts will be alike, and the *gamelan,* like a passage of its music, will be all variations on a single theme. The carved wood will be stained with red, and its designs overlaid with pure gold leaf. The carvers are not all members of the orchestra, but quite naturally they will do this work without pay, as good members of the community.

The *gamelan's* cost will be a thousand American dollars

and many, many days of labour. Near by, in the market place, the women of Kepisah are trading for the simple necessities of life, with coins worth one twelfth of an American cent.

V

ROOSEVELT swinging down the village street of Bengkel toward Kumis's home, like the pied piper with laughing children all about him, was a sight to behold. A tall man, a bit stooped but still athletic, with twinkling eyes behind thick spectacles, which looked with pleasure on the world. A perennial playboy. A grandfather who had never grown up.

His sandals kicked up the dust with a peculiar loping stride. A battered felt hat shielded his untrimmed thin hair. His shirt was open at the throat and cut at the elbows, and hairy knees bulged between the golf socks and the flapping shorts. His prominent nose was pushed a trifle to one side. He looked like his name, but had about him a devil-may-care shagginess that belied his features.

Over his shoulder swung his camera, like a troubadour's guitar. A battered old box, patched and peeling. Once he had dropped it in the sea, fished it out, sloshed it in a bucket of fresh water, and found that it would work. Once it had clattered down an endless flight of stairs in a Calcutta Raja's palace. It was temperamental, and he knew its whims like a wife's. Hold it horizontal, and its curtain shutter would slide at a certain speed. Hold it vertical, and there would be quite a new schedule of exposures.

Pictures—he was always making pictures. When, in Buleleng, he had developed a fresh roll, I would pounce upon it and marvel: every one a perfect exposure, a clear, fine image. Then, when it was dry, he would look it over, and dourly tear the negatives to pieces, one by one. Something wrong with the background, something wrong with the shadows. But here! this one is good! And this a beauty! Rejoicing! A successful day! Roosevelt was an artist.

What a renegade he was. He was doing his best to live down the family name's association with "the strenuous life." He had about him not one of those solid virtues which have made the West master of the world. He was incorrigibly irresponsible. He was invariably tardy. He was defiantly lazy. Business, the disciplined structure of industrialism and the modern social machine, these were anathema. Discipline? Bah! His proudest possession was a friendly farewell letter from his war-time colonel, which called him the worst damned soldier in the American army.

As a pure-bred American he would be inconceivable. His New York father, a first cousin of T. R., after founding the first telephone company in Europe retired to Paris, wed, and enjoyed the fruits of leisure. Born in Paris, the Franco-American son quaffed beer in Heidelberg with a student corps, sipped wine in Montparnasse with Picabia and Stravinsky, played tennis with his cousin at Oyster Bay, and in general enjoyed his patrimony. Then he grew rice in Texas, engineered in Missouri, barnstormed for the Wrights with one of their first planes, managed production for a motion picture company, did liaison in the war, broke his nose at the age of thirty-eight with the All-American Rugby team at the Pershing Games, invented what is now a nationally marketed product, lost it in the crash of 1920, and at last found congenial occupation as a French gentleman in Paris teaching retired American industrialists how to loaf. Chaperoning a tourist party through the East, he stumbled into Bali for three days. One look was enough. He took his party back to Singapore, shipped it home, and returned to Bali. There would be a hand to mouth living in guiding tourists now and then, and selling them photographs. And he would be in Eden. He would end his life there.

He was very happy, with no possessions to worry him. But he wanted a boat. He would talk of lecture tours sometimes, and motion pictures he might make, and the boat that this would give him. His voice would fly up to a high pitch of Gallic enthusiasm. His hands would flutter, flung out toward the Spice Islands and beyond.

"My boy, out there are thousands of islands! *Thousands* of them. No man knows how many. To many of them no white man has ever been. I'm going in my boat. And *you're* going with me, old cock!"

The Dutchmen, to whom a tour of duty in this quiet Bali was a boresome necessity, looked askance at Roosevelt. What? Did this man *want* to be here? Well, everyone knew that photographers were riffraff anyhow. And the man was lazy!

"Lazy? I *want* to be lazy!" cried Roosevelt. "Everybody with any sense wants to be lazy, but nearly everybody is afraid."

"Roosevelt wants to be lazy!" snickered these perplexed, stolid, industrious Dutchmen, who buttoned their collars high and tight. They drew aloof, as if to avoid infection, and passed by on the other side. And Roosevelt went on his way rejoicing.

You might have thought he would be lonely. But though the world knows almost nothing about Bali, faint hints have been passed about among the knowing. There was a steady trickle of civilized white men into the island. Painters, musicians, writers would drift there and somehow hang up their hats in Roosevelt's house. They would stay because they loved Bali, and also because they had found here a free soul and gay companion, who had retired to this sheltered haven and watched the world go by, amiably tolerant of all save prigs, salesmen, tightwads, prohibitionists, and drunkards. A man who had done what every man dreams of doing, and told the world in general to go to the devil.

Thus I had come to Bali, and to Bengkel, and Roosevelt had shown me joyous laziness, and adventure.

Perhaps, in calling this adventure, I had better explain. Adventure, it seems, is something shot through with high-speed action, something fearful or unpleasant. It may have pleasant connotations, as when the fierce tribal chief presents to you his beautiful daughter, but fundamentally

Adventure is hunger at sea or thirst in the desert. The world abounds in Adventure. You go into the disease-infested, mosquito laden jungle—without witnesses—and shoot a tiger at close range, feel the wind from poison darts of Dyak blow-pipes, knock off with a cudgel the cobra which has coiled about your leg. You have a terrific fever; and if you are not writing for the *Ladies' Home Journal* perhaps the delirium clothes in golden memory some damsel in a seaport pleasure house. If you are especially fortunate, you may be shipwrecked and swim three miles to a lightship with a pussycat on your back.

Something is wrong with me. Probably I am not a man of action. Probably I have been too discreet about courting discomfort. Still, I have been around a bit. I have worked in a half dozen ship's forecastles, but never encountered a good he-fighting episode. Once fever struck me on a mountain top, with none at hand but a Spanish-speaking brown crone; but instead of glamorous delirium all I recall is the vague unpleasantness of walking twenty miles with chills and headache. The Sultan of Bima took me hunting in Sumbawa, with three hundred naked, yelling spearmen to beat up the bush, and all we got was an old wild sow too tough to eat, and unclean anyway to my Moslem host. I have wandered at night through the dark native quarters of Singapore, and my only trouble was to make my rickshaw boy resist the importunities of Chinese pimps. I have travelled thirty thousand miles and more, and slept in outlandish places, and never encountered so much as one bedbug.

The luck which follows men of action, chapter after

chapter, eludes me. One writer had his camp in West Bali trampled by a herd of wild elephants. I could find no one who had ever seen an elephant in Bali. There are no poisonous snakes there, either, and tigers are very scarce and timid. The people are not unfriendly, not fierce, much less dangerous. Where then adventure?

But these days in Bengkel were adventure, of the most authentic sort. Emotional, psychic adventure, if you will. I was the young man in Balderston's *Berkeley Square,* suddenly finding himself in the eighteenth century. I was the Connecticut Yankee at King Arthur's Court. I was a cog from a western machine suddenly ejected into another world, ushered by a man who had discarded western valuations into the life of people who never knew those valuations. I was before the industrial revolution, before the Christian era. Vaguely dissatisfied with my niche in the rolling mill, I had set out without aim to wander, hunting for something, I knew not what. I had loitered along the ways the white man has made for himself, a tourist like a zoo visitor, staring beyond the bars at the quaint ways of queer peoples. Now suddenly I found myself on the inside, a part of something rich and strange; and far away was a vision of the West, like a faint memory of a glittering Woolworth store window.

Of course we could never reach a status of perfect, intimate understanding with these people. We were white *tuans.* Roosevelt was *Tuan Besar,* the great lord; I was *Tuan Muda,* the young lord. Our predecessors had established the White Man's Prestige.

In these days Roosevelt had a recalcitrant cut on his

shin, whose tropical persistence required frequent bandaging. Nothing would do, then, but that every boy with a scuffed knee about the village must be brought to our *balé* to share the dressing. I soon became a part of this impromptu out-patient clinic. One day when we arrived in the village I noticed that little Runis, the dancing girl, who was nine, had a badly blood-shot eye. I had some argyrol given me by an oculist in Manila, so I told her to come around and I would do something for it. Three times a day for the next six days, without being told, that solemn child would come to our *balé,* and I would drop the medicine in her eye. She would hold her head back carefully, to keep the drops between the lids. We got to be good friends.

Other eye patients began coming. There was one old woman whose body was wasted away, whose teeth were gone, who eyes were sore and red. She came every day. One day she brought a chicken that she wanted to give me. There was an old man, too, whose left eye was blind. But I could not help him. There were not very many of these people. Contrasted with the pitiful sights I had seen all through the East elsewhere, they served but to emphasize the glowing health of all this island.

Thus we were breaking through prestige to friendliness. There was the barrier of language also. Roosevelt had broken that down pretty well, for himself. Roosevelt was a daily reminder of a feeling of linguistic inferiority that had been bothering me for some time. Every educated, not to mention uneducated, European I had met spoke several languages. Arrogantly the ignorant English-speaker may go almost everywhere, sure others will know his tongue. But every week that I was with him, Roosevelt used four languages. And gradually I was realizing that most of these brown folk around us spoke three languages. Among themselves they spoke Bali, a difficult tongue which very few white men know. If they had been speaking in the presence of a Raja or other royal person, they would have spoken High Bali, which is quite another language. With us, or at least with Roosevelt, they spoke Malay, which is the common tongue of all the myriad nations in the Indies.

I was learning to speak a little. In Borneo, Singapore, Java, I had picked up a few words. Now I would converse stumblingly with Kumis, and he would tell me the names of things. The simplicity of Malay charmed me. Its idiom is more like pidgin English than anything else in my experience. It is quite different from learning a new European language, in which the basic grammar is familiar. Malay is a language without inflections. To make a noun plural you say it twice. The word *makan* means "food." It also is the verb "to eat" in all its moods and tenses, and its meaning extends beyond eating to any form of consumption. If you take a stroll in the evening, you "eat the air." If you indulge in dalliance, you "eat a woman." It is astonishing how many meanings a single word can convey. *Suda* means "finished." Usually it merely puts a verb in the past tense, but it may express anything from the binding of an agreement to the breaking of a friendship.

This economy of vocabulary makes Malay a racy, colourful tongue. When I learned that the sun was *mata hari,* "eye of day," I was charmed. I enjoyed thinking of a fried egg as *mata sapi,* "eye of a cow." I was delighted by the idea that pink was "young red." But when I was told that *prumpuan,* "woman," meant literally "something to be forgiven," I knew that it was fun to learn Malay. And as I learned words, I grew closer to the life around me.

There was the barrier, too, of a vague feeling engendered by months of hotel habit that anything "native" was likely not to be clean. But this soon vanished. These people were very clean. Every morning in Bengkel Roosevelt and I bathed out by the well, between the compound wall and the back of Kumis's house. Madé would draw a big basinful of water, and we would plunge it over ourselves with a coconut shell dipper. This was just the sort of bath we would have had in the best hotels, from Singapore on through the islands. Our native friends bathed every day too, in a stream that ran through the bamboos a hundred yards away. Its water was brown, from the unpolluted earth of the hillsides. Its banks rose sharply six feet, and bamboos shaded it. The girls kept their *kains* about them, or carefully lifted them over their heads as they sank into the water. The men took their *kains* off, and concealed themselves frankly with one hand. There was a broad, smooth boulder there, against which they would scrub their backs. You hardly ever pass a stream in Bali where there is no one bathing. Every spring in the island which has possibilities has been bricked up and diverted into fountains. Most of these bathing places have concealing walls and separate places for men and women, but some of them have not. There are strict ethics at such places. A few years ago in North Bali a young man so far forgot himself as to paddle a bathing girl. Angered, she appealed to the old native law, and the impudent youth was sent to jail for three months. But I don't think the girls are generally so priggish.

There was also the vague uneasiness engendered by the locks Dutchmen put on their property, and by stories of the innocently light-fingered ways of tropic islanders. That soon vanished. My suitcase always lay open there in our *balé*. My things were likely to be left lying around the place, garments and toilet implements which were strange and interesting to these people. They were always carefully put away for me. I had money in my suitcase, and batik *sarongs,* which were very much prized in Bengkel. The whole village flowed in and out, and I never lost anything. At home I never can keep a finger nail file. I don't know where the confounded things go, but they disappear. My file was a very interesting novelty in Bengkel. When your Balinese has inclination to be foppish, and show an aristocratic lack of concern with manual labor, he lets the finger nails of his left hand grow, until they are perhaps two inches long. So the idea of manicuring the nails took hold quickly. It became a village fad. People were always dropping in to use that file. I still have it.

Anyone reared in America is conscious of the colour line. I am. One cannot help being so. But fortunately the American colour line is so conveniently specific that it need not be argued away in Bali. The same Southerner who segregates an octoroon will boast of his descent from Pocahontas and vote for Charles Curtis for Vice Presi-

dent of the United States. Well, the people of Bengkel
are not Negroes either. I had a time at first, trying to
identify in them negroid characteristics, but soon had to
discard all ideas of the Congo and Mumbo-Jumbo. Africa
was a long distance away. These were Asians, of Malay
stock, with an infusion of Aryan blood from India. I had
to jettison, too, my whole mental cargo of tropical island
lore. Though there is a faint infiltration of Polynesian
blood, which is dominant in the eastern Indies, Bali is not
a South Sea Island. Its people are not South Sea Islanders,
and have very little in common with them.

So, gradually, came my orientation in Bengkel. I came
to feel at home with these people. But never did I become
accustomed to their personal beauty. Again and again during the day it smote me, in effervescent surprise.

Especially the young women were beautiful. Kumis's
niece, Renang, was just coming into womanhood. A dainty,
golden thing she was, with features like an orchid, with
skin that glowed softly beneath the tropic sun. Like all
the other girls, she wore but a *kain* or *sarong*, a wide
length of cloth wrapped about the body and belted at the
waist. It was white, with gay peacocks of red and blue.
Renang's ears had been pierced when she was a baby and
stretched with plugs of gradually increasing size, and
through their lobes she now wore rolls of white palm leaf,
an inch in diameter.

She had long silky hair. Most of the hair in Bengkel
was black, but hers was a rich dark brown. She perfumed
it with frangipani. She paid a lot of attention to it. She
never used any hairpins. She had several ways of doing up

her hair, which I can't describe. Usually it was wrapped around the head and interwound with a scarf, with the ends hanging in a feathery fringe. Sometimes she would knot it in back, with a tassel dangling. Sometimes she would roll it inside itself in a great puff on the side of the head, tied in place with a few strands that had been separated from the rest.

The girls would tend to each other's hair. Somehow, sitting together, they grouped themselves naturally in poses like a sculptor's dream. I never saw them use a comb. But always when some of them were sitting together, the hair of one of them would come down, and the rest would finger it and separate the strands. Clean bright scalps shone through the parted locks. They would hunt through it carefully for any of those little animals they called *kutus*. They seemed to find very few. There could be little game in territory so consistently hunted. They hunted for the sport, rather than for the quarry. Though the custom undoubtedly originated as a necessity for people who insisted on keeping clean and did not have the chemical means of doing so, this co-operative hunt had practically become the social equivalent of an afternoon sewing circle. It was a gay, gossipy ceremony, and altogether charming.

And now, while some feminine readers squirm squeamishly, I must insist that these people of Bengkel were personally quite as meticulous as our own great grandparents, and undoubtedly Renang was more so than Paolo's Francesca or Ivanhoe's Rowena. There are virtues in the fastidiousness of this Age of Listerine, but let's not be finicky.

It is well to be specific about attitudes. I shall be accused of writing about Bali as a romantic, without the excuse of being Irish. But I think I have made clear that I reached Bali in a very disillusioned frame of mind, for my eyes persisted in seeing things. I kept the same eyes in Bali, and I have not ventured into the world of Let's Pretend.

Eyes differ. When Harry Hervey saw the dancers of Angkor, he saw the Apsarases of Cambodian legend in a poetry of motion. When Harry A. Franck saw them, he observed that they had dirty necks.

My capitulation to Bali did not come about completely in a day or week. Charming though it be, Bali is no saccarine Utopia, monotonous with felicity. As in other tropic countries, milk and honey come in cans. Men and women grow old and shrink to hideous phantoms of themselves. But even more than in such matter of fact affairs, it took time to focus the eyes. Consciously or unconsciously, like every American whose life had been bounded by two oceans and who is not one of an esoteric group, I had a Credo. It might be more or less approximated as follows:

Any coloured person is in all respects inferior.

A beautiful woman is one with face and ankles like those in hosiery advertisements.

A beautiful skin is white, like mine.

Religion is Christianity.

Beautiful dancing is what you see in Mr. Ziegfeld's shows.

Beautiful music is that made by Paul Whiteman or (theoretically) Toscanini.

Art is something queer men make for magazine covers and museums.

The object of life is to "get somewhere."

So long as Bengkel was seen from the standpoint of these and a myriad of similar definitions, it was nothing but a parade of curiosities. There were curious and amusing things in Bali. But I saw something more. I stayed in this *banjar* of Bengkel long enough to get used to it. I looked back over ten thousand miles of sea and saw those distant, strange United States.

These people about me in the home of Kumis had feet, ankles, hands of the finest daintiness; their limbs and bodies were gently modelled; their every movement was instinct with grace; and in the depths of their great eyes glowed a peaceful tenderness. Their noses and lips were different from ours. Their skins were brown. And now I declare that, in this environment at least, under the tropic sun, brown skins are far more beautiful than white skins ever can be. Sometimes, caught in a wave of kindred feeling, I had an impulse to strip off my shirt. I quickly covered myself. I was so indecently white!

One's eyes adjust themselves. Sometimes we would encounter parties of tourists. How pasty-faced they were! How long their noses!

One day, on a ship westward bound, I was telling these things to a young woman who had been visting in Malaya. Her hair had been waved by an expert *coiffeur;* her nails were fastidiously polished; Paris had cultivated the charm and precision of her dress. Every device that little Renang knew not, she had used to enhance her natural beauty;

though being British, she walked like an Englishwoman.

"But why," she exclaimed with a lady-like shudder, "didn't you live in a bungalow, with servants? I never heard of anybody wanting to live with natives."

So, after paying what tribute was due to her own lily whiteness, I told her things about the people of Bengkel: that they had developed a civilization which was not one of germicides and bathroom fixtures; that they were far advanced in some respects in which we were generally barbarians; that they were beautiful— She interrupted me.

"But if you married one of them, you'd soon get tired of her!"

VI

DAY after day Renang sat weaving. She was on the little shady platform under the high-peaked granary. The light shone through from behind her, etching her delicate profile and the soft, gentle swelling of her bosom. She was there just a few steps from our *balé,* and I liked to sit and watch her. Sometimes she smiled at me as she walked through the yard.

I used to sit beside her to see the deftness of her fingers. She lifted strings to separate the threads of warp, tossed the bobbin through, then jammed the thread down solidly with a long slim blade of wood. Renang had been a *legong* dancer and her hands were very swift. The warp stretched from a curving wooden yoke behind her waist. She was weaving a plaid design of green and yellow. Two widths sewed together would make a pretty *kain.*

Sometimes she would smile at me. She did not chew betel, and her teeth were white. I would admire the slimness of her hands, by signs, not touching them. She seemed to like that. She could not speak much Malay, and neither could I, for that matter. So we smiled at each other. That seemed to be enough.

I thought of Renang sometimes in the evening. Up beyond the brook-banks near Kumis's house were moon-drenched lanes among the bamboos, beckoning to strollers. But there was no loitering there. The lovely night lay virgin in her lonely scented bowers, vaulted with Diana's silver; and not a lover knew her. It would have been pleasant to see men and maidens in the dim light, pleasant to have lingered, not alone. I thought of Renang sometimes in the evening. But I did not see her then.

Sometimes when other girls were in the house, Renang would join their merry group. They would cluster around me, laughing and jostling, and I would have to give them cigarettes. There never were enough cigarettes to satisfy them. But I never saw Renang alone.

It seemed as though Renang and I would never get beyond the smiling, weaving stage, until one day we all decided to go bathing in the sea. Somehow as we started Renang was sitting on my lap in the front seat. That may have been because she was invited after the back seat was filled. It is not good to be too diffident.

We were all piled into a little battered Chevrolet, and very merry, seventeen of us, big and little. We drove slowly down four miles, to bathe by the coral temple.

I was polite to Renang. Of course, she knew I liked to have her sitting in my lap. Surely, I should have been remiss if she had not known it. But my behaviour on this ride was of almost severe propriety. I have met many virgins who would have thought they were being slighted, had I treated them that way on first acquaintance. But all the while as we rode toward the sea Renang grew shyer.

A handsome, graceful young man was coming up from the sea when we reached the end of the road. There was a fisherman on the shore, casting his net. The youth had bought some little fish there, and was walking home with them. When we got out of the car, Renang ran over to speak a minute to him, then trotted on to catch the other girls, walking down the beach.

The girls went fifty yards down the sea shore before they entered the water. Some of them just waded. Some of them took their *kains* off after they were in the sea. They did not stay in very long. They were not swimmers, and I think they had come just for the ride.

René, our cook, did not go in the water at all. She went and bought herself some fish. They were little white fish three inches long. When we got back to Bengkel she was cooking some for herself, and she showed me the big batch she got for five cents, as an object lesson in thrift.

Roosevelt and I did not stay in the sea very long either. The equatorial ocean is listless, enervated. The tropic air seemed chill as we came from the brine.

We returned to the car to find the back seat filled and running over, and there was Renang in it. The only one left to sit in my lap riding back was dear homely Madé, wife of my friend Kumis. I did not mind holding Madé on my lap, but there was a difference.

The next morning I glimpsed Renang walking through the door yard with a water pot on her head. She did not weave at all that day.

The following day she was back there, weaving; but she did not smile at me at all. She did not look at me.

I had been, you may have surmised, right royally snubbed.

The mystery of Renang's nocturnal whereabouts did not long remain unsolved.

Almost every night from deep amongst the compounds came the rattle of drums and cymbals, the whine of a one-string fiddle. What might it be, this noisy night life?

"*Janger sekolah,*" said Réné the cook.

The *janger,* to be sure. I had seen the *janger.* But the school? Babad would take me there, down through the starlit, phantom-laden night.

In through a narrow dark alley-way we went, until we came to a place where a group of women blocked the way. Two pretty little girls were sitting there, with tiny oil lamps, selling peanuts and cigarettes. From beyond the crowd came a sound of singing, shouting, and the syncopated rattle of drums. We elbowed through.

We came into a little square courtyard, with *balés* all around, and seated cross-legged on mats in a hollow square, girls facing girls, boys facing boys, danced the young folk of the village. Drums, cymbals, fiddle struck off an itching syncopation, the plump little girl struck up a lilting nasal tune, and the whole crowd was off in a burst of noisy rhythmic fun.

The boys seized the burden of the dance. Hands on knees and elbows high, they jerked their torsos; now their hands flew, like a Harlem belle doing the Charleston. Now they were on their feet, like college cheer leaders. They scorned to sing, gruff shouts in unison were their part. "Ka *JA*—ka *ja ja; ka JA*—ka *ja ja!*" and then in hoarse whispers, "Go *wé*—go *wé*— Go *wé*—go *wé*—" They leaped upon each other's shoulders in balanced pyramids, and now they were seated on the ground again, and the little girls were singing, *"Janger, janger, janger!"* "Bean porridge hot, bean porridge cold." That was about as much sense as it all had. It was a wild dance, a jazz dance, and all the dancers had a wonderful time.

Now there was something about this dance which seemed impossible, even as I watched it. For while their necks were perfectly still, the gay little heads of these girls were sliding from side to side, without turning or nodding in the least, jerking from side to side in the primitive pounding rhythm of the drums. And there in the centre of a row of seven, eyes snapping, head jerking, hands fluttering, sat Renang. So this was where she spent her evenings!

I had seen her dance before, but had not known her. Then she had worn a great spraying crown, and her brow was painted. But now they did not have their costumes. For this evening practice the girls had bound their breasts with scarves, and that was all.

When Renang and Rupag were retired, at the age of twelve, from the dancing of the classical *legong,* the normal thing for them to do was to settle down, tend to their weaving, and wait until time for getting married. But just at that time a fad was sweeping Bali. Spontaneously appearing in the North, a new dance seized the fancy of Balinese youth, and soon clubs of boys and girls had been organized in all parts of the island to dance the *janger*. It did not last long in some vicinities, but here in the villages around Den Pasar it held its popularity; and at Bengkel it seized a place of predominance in the village life. Renang and Rupag were the leaders in it.

When I first saw the *janger*, I suspected that it had been concocted for the benefit of those few tourists who straggle into the island, but I soon realized that this was an error. Doubtless the occasional visits of travellers give it a certain stimulus, for they add to the revenues received from native festivals at which the club performs. But there was nothing commercial about it. Every penny went into the communal fund of the organization, and if any dancer was seen receiving a present from a visitor, that too went into the common purse to provide costumes. Réné, our cook, was the *janger's* treasurer, and she guarded its funds with a watchful eye.

Almost every night these young people at the *"janger* school" would work out new ideas, practice new songs, for the dance was always changing. There hadn't been a wedding in Bengkel for a long time, for most of the village's eligibles were in the club, and married persons could not belong to it. Night after night I used to watch them. This rollicking, popular dance, with its roistering, syncopated rhythms, its swaying seated dancers, dancing with arms and hands and shoulders, was much in contrast to the fastidious precision of the classical *legong*. But it had the joyous grace of spontaneity. And especially it interested me as evidence of a nation's reaction to foreign things.

One frequently finds in the East, depressingly, that natives are imitative, seizing especially on American dress, jazz, and dancing. The Balinese does not imitate, but when confronted with something strange, observes it and turns it into something peculiarly his own. The artist, as we shall see, has done this in the temples; and the young people have done it in this dance.

This struck me especially when I looked at the costumes, the first night I saw the *janger,* costumes which were whimsically discarded a few weeks later in favour of more purely Balinese things. The girls, in their traditional Balinese crowns of spraying flowers, had donned the Japanese *baju,* shirt, lightened with a shoulder sash of bright colour. Then they had looked upon the tin cans which they scorned as water containers, and observed that they were bright. So from them they made wide silver girdles The youths had put on fezzes, burlesque moustaches and whiskers, and Solomon Levi gestures to satirize the Arab traders of the towns. Ordinarily scorning shirts, they had donned white ones with collars attached, and four-in-hand ties, all topped off with gaudy epaulets and hanging broad collars of tinsel and glittering mirrors. Incongruous, of course, but intentionally so; the Balinese are not blind to the movies and occasional tent shows which drift into the northern towns. Once in a holiday parade in North Bali, I saw two American Indians and a black-face who might have been either Mr. Moran or Mr. Mack.

Many of the rhythms of the *janger* doubtless filtered through from Tinpan Alley, by way of the Malay Opera which tours east from Singapore. Many of the men's gestures were plainly from the Malay Opera, and their leaping pyramids probably had come from some visiting troupe of Chinese tumblers. The collegiate cheering of the men, I am convinced, came from the retching klaxon horns of motor cars. It is possible, of course, that the native chauffeurs imitate the rhythms of the *janger,* but at any rate the two are identical.

This *janger,* when danced for spectators, would rollick on for an hour and a half, interspersed with solo dances. It would end in a bit of masked drama, with fierce-faced, long-tailed monkey gods leaping about, with the fleet-winged Garuda bird, and Renang with a bow and arrow slaying it. Then would dance the Rangda, torch-eyed, and terrible, with hungry mouth and trailing, uncouth hair.

Among the many dances of Bali, the mask is almost ubiquitous. There is one dance, the *topeng,* in which the artist for hours performs in dancing, monologue, and pantomime, all the while changing from one mask to another. The masks are cunningly contrived of wood, paint, and hair—comic, solemn, fierce, horrible. The most horrible and ubiquitous of all is the Rangda.

Rangda was a widow. Rangda was a bad woman, they will tell you. Some also will tell you that Rangda was the favoured wife of the Raja Deri, who saw another beautiful woman and took her, which made the Rangda angry. The Raja and his new wife died, so the story goes; and the people blamed the Rangda and sought to kill her, though she was innocent. Whereupon the gods saved the Rangda and made her a spirit, as which she scourged her persecutors in terrible revenge. That is one story, but it is my belief that the Rangda is a memory of days when a Hindu widow must burn herself on her husband's pyre, and a surviving widow was a terrible thing. Even today, in remote parts of Bali, all widows are called *rangdas*. But the Rangda is even older than that, I believe. She is a pre-Hindu vision of evil, source of aches and ills, man's conception of the wrathful aspect of God. In Bali's drama, when the great god Shiva becomes angry, he retires and reappears, incarnated as the Rangda.

The Rangda mask is very sacred. It is kept in the temple. When someone is sick, it is sprinkled with holy water.

I liked to watch the *janger* itself, but especially it was fun to sit there in a *balé* during rehearsal. Nearly everyone was there, an amiable, sociable crowd. If for no other reason, I should find the Balinese kindred souls because they know the best time of day is after sunset. And if for no other reason, I should admire them for their ability to go without sleep. They practise staying awake when they are children. This night club was thronged with little boys and girls, waiting for nothing so much as to grow big enough to join the *janger*.

As I sat there swaying with the rhythm of the dance, there came upon me an incredulous amazement. There was always a peculiar lack of demonstrativeness among these people, a peculiar lack of physical radiation. It was hard to imagine them in any intimacy. Renang and the other girls of Bengkel resembled in figure and dress the Venus de Milo, and they were just about as coquettish and provocative as that awesome piece of marble. But here was something more than that; a traditional belief, a literary invariable, was being shattered. Here was a night-time dancing club, brown-skinned boys and girls of mating age, swaying to the pounding of the drums. By all expectations it should have been an orgiastic apotheosis of the amorous night. But in it all, singing, swaying, leaping, was not one jot of sex.

Aside from a rather prim personal modesty, there was never a sign of sex consciousness in Bengkel. Never an overt act of mating. Never a caress or a seductive glance. Never in dancing, music, art, any hint of that mysterious, fascinating thing that we call sex. True the phallus of the god Shiva is his symbol, and some temple carvings have the most amazing genitals. This very bluntness clinches the point. Adam ate the apple before he mutilated the fig tree.

All this that we make so much fuss about was so simple here. The quest of food and the quest of a mate are plainly the most fundamental things in life. They are the source of most of the world's effort, and most of the world's woes; and we of the West, with our cook books and our romantic novels, have contrived to complicate them mightily. In Bengkel there was not much trouble about getting food, and it was taken without much ceremony when one was hungry. Mating was a simple matter also.

With these needs out of the way, the man of Bengkel turned his attention to more interesting things. Eating and drinking were important, certainly. But why get excited about them? Why let them dominate the mind?

These thoughts ran through my head as I sat there watching the *janger* practice. I knew perfectly well that there were lovers among this dancing group, and probably (the climate being what it is) assiduous ones. But from what I saw I never would have guessed it.

There was a breeze this evening, and a single oil lamp lighted the courtyard with fickle flashes. On a little flight of steps, beneath the lamp, I saw that handsome young man whom Renang had greeted that morning by the sea. The light glinted on the high rakish brow that sloped back to his unruly shock of hair. What an intelligent, good-humoured face he had, what easy, graceful motions!

Who was he? I asked. He was Ida Bagus Gidéh, the priest's son. When his father died, Ida Bagus Gidéh would be the *Padanda*, the leading man of all this neighbourhood.

The priest's son saw me looking at him, and flashed a friendly smile. What a romantic-looking fellow he was. He had a glint in his eye that made me imagine he must be quite a devil among the damsels of Bengkel—young lord of the manor and all that. It must be pretty near time for him to be getting married. He would have a pretty girl.

Renang, perhaps? Maybe it was Ida Bagus Gidéh that

Renang liked. I remembered that she had gone to speak to him that day beside the sea. I looked at Renang, dancing there, head twitching, fingers flying, all unconscious of my gaze. What a pair they would make. As I have said, there was no sex-obsession among these brown folk. But I was white. My romantic imagination would not down.

On a mat at the centre of this hollow square of two dozen dancers, sat a good-looking young man named Nonga, dancing all alone. With arms, fingers, eyes preening in fantastic motion, he made an axis for it all. When first you saw him, in these postures of the ancient *gandrung* dance, he seemed a little wonderful. But after a while you realized he was not so very good. He kept repeating himself. He even seemed a bit ridiculous, preening there.

Now I noticed he was hesitating, fumbling, trying something new. His eyes were turned aside. He was watching Ida Bagus Gidéh.

Ida Bagus Gidéh was dancing, giving lessons. Incomprehensible as the *gamelan's* music, the motions of this young Brahama. He sat cross-legged, posturing. Arms extended, arms bent, fingers strangely grouped. The face a mask, majestic. Motions slow and sudden, in balanced patterns. There was a fastidious aestheticism here that defied words, defied thoughts, that set in vibration unsuspected emotions. It caught me, held me, with a breathlessness I could not understand. Gidéh was become the impersonal personification of something beyond dreams.

All the while the *janger* went on dancing, singing its gay and lilting minor tune. My eyes wandered to Renang. She was watching Ida Bagus Gidéh.

The lamplight flickered down from above his head, making a flashing bright sequence of angular highlights, invisible shadows. It etched his profile, flashed upon his eyes, lightly caressed the rippling muscles of his torso. His arms were snakes, not writhing; they reared and struck with the kingly swiftness of the cobra.

And now, as suddenly as he had begun, he stopped. This majestic deification of line and form had become, all at once, a familiar, genial smiling youth.

I knew not how long this dancing might go on, but soon I knew that it was time for me to go to bed. *I* couldn't sit up all night and be fresh in the morning, even if little Babad was sitting there as wide awake as any. Sleepily I strolled to bed.

Incorrigibly romantic thoughts ran through my head. Not a thing had justified them, but they persisted. Ida Bagus Gidéh and Renang, Renang and Ida Bagus Gidéh. Was Ida Bagus Gidéh the man Renang had chosen?

I had no idea what Renang might be doing when the dance was over. Renang slept in a *balé* pretty much like mine. Not until after marriage would she rest within closed doors.

VII

IT was during the rice harvest, some fifteen years ago, that Réntung and Kumis worked together in the fields and knew that they were mates. She saw Kumis, a stalwart youth of twenty, striding with two golden sheaves of paddy balanced from a springy stick across his bulging shoulder. He saw Réntung, a pretty girl of eighteen, walking carelessly with a great bunch of feathery grain upon her head. He had not mingled much with girls, nor she with boys.

It was not romantic love that sprang between them. In their tongue there was no word of love. Between Réntung and Kumis there was something much more simple, much less constrained, than that. They looked upon one another, and each saw that the other was good.

(Perhaps they had looked upon each other as did Ida Bagus Gidéh and Renang, I thought in these days.)

Kumis plucked a great red flower and wore it over his ear. The flower said: "I have seen my mate, and let no man deny me."

So it came about that all the village knew a secret Kumis was going to marry Réntung.

At a certain hour of a certain day, Rentung's parents were away from home; and Kumis's friends went there and took Réntung away. Almost immediately her parents returned, and when they found that she was gone they raised a fearful row. At the same instant the whole village discovered the abduction-elopement, and rushed to Rentung's home to condole with her parents, who were full of grief. Then the parents rushed about the town, searching for Réntung, and the village helped them with the hunt. The village bell beat furiousiy. If they had found Réntung they could have taken her home. They searched everywhere, except in the house of Kumis's father. That would have been against the rules.

The next day two of Kumis's friends called on the bride's outraged father. They explained that Kumis meant to be a good son-in-law, and had not intended any offence. Then Kumis's father called on Réntung's father, and suggested that they should continue to be friends. It was a good thing that their families should be united, anyway. Kumis's father offered a present; thirty dollars, I believe. So then everything was settled.

On the third day Kumis and Réntung appeared again in the village, and there were great preparations for them. Everybody in Bengkel had brought food to the *balé banjar*. The *legong* danced, the *gamelan* played, and the old priest sprinkled holy water and flower petals while lacy palm leaf offerings were placed before him. There were roast pigs, rice cakes, and great bamboos of palm wine. All Bengkel made festival, through the day and night.

It is too bad for this story that Kumis did not live a few miles away, for then their marriage would have been livelier. There would have been a thrilling abduction, and Réntung would have been dragged down the street, fighting like mad against her captors.

Not long ago a young Dutch *Controleur,* recently arrived with his young bride, called the Pungawa to him.

"I saw a terrible thing today," he said. "As I was driving in broad daylight, I saw a young girl being dragged along the road by three men, though she was resisting furiously. Such things must not be allowed. I stopped them, but I must count on you to prevent such things."

"Tuan," said the Pungawa, "you have stopped a perfectly good wedding."

In consternation the chivalrous Dutchman ordered that things be set straight immediately. But it was not so simple. For the girl's parents were found to be elated. The *tuan* had stopped the wedding, had he not? Surely, therefore, the *tuan* wanted the girl for himself. Her family was to be greatly honoured.

But while embarrassment was thus heaping itself upon the young Dutch husband, the girl settled the matter herself. For she had already chosen her man, and did not want the *tuan;* and among this people a girl marries only the man who pleases her.

Having married Kumis, Réntung was responsible for the work of the household. But her lot was not hard. There was not much work to do; and if the work increased it was a good chance that there would be another wife to help her do it. She didn't have to fetch her husband's slippers, because he had none. She had no dishes to wash nor table to wait on. Even if she had had a baby, there would have been no baby clothes to launder. And if she did the trading in the market place, her western sister did the same. She got a lot of enjoyment out of life.

And there wasn't any romantic passion to upset things. When I say there is no love in Bali, the rather large number of men who have been stabbed for stealing wives might be taken as contrary evidence; but those who know the people best agree that these were not crimes of jealousy, but matters of pride and resentment over loss of valuable property. This might seem to put woman back on the ethical plane of the Ten Commandments, which rated her as a chattel along with her husband's ox and ass. I suppose it does.

But Réntung as a Balinese woman had a position of far more relative dignity than that of most oriental women. In Japan, where they do have a conception of romantic love, a poor father may seal his daughter for a term of service in a brothel, and hers will be considered a laudable aid to her parent. In India, a father marries off his daughter when she is a little child, or he may send her to be a temple prostitute. It is commonplace that oriental brides frequently are strangers to their bridegrooms. In Bali a girl chooses her own man. Rape is a serious crime.

Prostitution was unknown until foreigners made a demand for it. It is now unknown in the villages.

Under the old Balinese system of law and courts, which still operates with slight regulation from the Dutch, Rentung's status was somewhat defined. If during the first year of their marriage Kumis had had cause for displeasure, he might have sent Réntung home and got his wedding money back. But if, after living with him for five years, Réntung had been able to prove in court that he had been cruel to her, she might have obtained not only a divorce but also one third of his property. If she had run away from him, she would have owed a $16 fine to the Government and $10 to Kumis. I am not in a position to say that these laws are any more respected than our own divorce laws. My impression is that if Kumis and Réntung had desired to part company they would have been inclined to handle the matter for themselves, without formality. But at least the laws were there, engraved in their palm leaf books.

Anyway, Kumis and Réntung were married and lived happily, but had no children. After a few years, since Kumis was a man of property with a considerable household to keep up, and since he wanted children, he married Madé, who made him a good wife also. He married a third wife, too; but she was lazy. So Kumis divorced her, and got a refund.

VIII

THE rain sang a droning song, and the eaves of our *balé* spattered arpeggios in the puddled yard.

We were very cool and snug, there in our *bale,* Kumis, Roosevelt, Ida Bagus Gidéh, and I. Opposite, under the shelter of the granary, Renang sat weaving.

I had got to be quite good friends with Ida Bagus Gidéh. It had surprised me at first to learn that this genial young gentleman was a Brahmana. Only his good looks and graceful manners distinguished him from the crowd. A Hindu of priestly caste, I knew, would have nothing to do with common men. He must stand aloof, preserved from contamination. But Bengkel was not Calcutta. Ida Bagus Gidéh mingled in all the life of Bengkel. He was very popular. He was respected too. The people always spoke his name in full.

This caste system, with four airtight compartments of society, which somehow paradoxically became in general practice a virtual social democracy, had puzzled me, especially because my host Kumis appeared to have no definite understanding of it. He said he was a Satria, which is the caste of royalty, but his neighbours said he was a Sudra, at the bottom of the scale with nearly all the other villagers. Ida Bagus Gidéh finally had explained that. "Kumis's fathers long ago were Rajas of Gianyar," he said. "But the Raja finally was beaten in war and driven out. He lost caste, came here to live. So Kumis is a Sudra."

Anyway, on this particular afternoon Ida Bagus Gidéh was sitting by our table, with green cloth spread out before him. He was measuring a pattern on it with a ruler, and marking it with a pencil.

"What are you making?" I asked.

"New costumes for the *legong* dancers," said Ida Bagus Gidéh.

Kumis was by the edge of the *balé* floor. He had no need of a chair. He squatted down with his feet flat on the floor and his knees against his chest. His right arm was straight, resting on his knee, and extending forward as a balance weight. He was perfectly comfortable.

Kumis got up, helped himself from the betel box that was always on our table. With the bronze bird that was a knife he sliced a bit from an areca palm nut. He took a heart-shaped green *siri* leaf, rolled the nut in it, dabbed it with a bit of white lime made from coral, put it in his mouth. He took a wad of scraggly scrap tobacco, rolled it tight, and put it between his lower lip and teeth. He squatted down again. He looked very stupid and contented there, with the tobacco wad protruding.

They chew *siri* all through the islands. It stains the teeth the hue of ebony. The Moros, up in the Philippines, smile at you horribly, from what seem toothless bleeding mouths. Kumis's tobacco kept his lips unstained, and kept his front teeth clean. Kumis had a fine white smile. The girls of Bengkel chewed *siri* too, some of them, but they had white smiles. They made a stain out of coconut husks, to make their molars an even black. But they polished their front teeth with brick dust, and had white smiles.

Kumis was squatting there contentedly. After a while he took the tobacco from his lips. His mouth ejected a stream. It made a spot of living crimson in the rain-stippled yard.

I tried chewing *siri* too, nut, leaf and lime. I didn't take any tobacco. My tongue all went prickly with a piquant taste like wintergreen. I spat a little. It wasn't red, and I wondered what was wrong. The leaf and nut all went to tiny bits. *Siri* wasn't at all easy to manage, like chewing gum. Suddenly my mouth inside was like a gushing fountain. I rubbed a hand across my lips to dry them. Its palm was streaked with crimson.

Pretty soon I got rid of the *siri* and took a cigarette. We always had cigarettes on the table, beside the betel box. We smoked Thomas Bear's Elephant Cigarettes in those days. They came ten in a little paper package, with a fine green sunrise outside and an animal picture inside. They cost two American cents a pack. Really, they were very good cigarettes, when you consider how many friends we had to smoke them.

Ida Bagus Gidéh took a cigarette too. We were having a very pleasant time sitting there in the *balé* while it rained. Sometimes it rained so hard that the eaves dropped a silver curtain, and dimly through it I could see Renang weaving.

Kumis and Ida Bagus Gidéh were talking about something. They were talking in Bali, so we couldn't understand. A certain tensity grew between them. Back and forth, back and forth they talked. They might be arguing.

"What are you two talking about?" asked Roosevelt after a while.

"My white rooster can beat his brown one," said Ida Bagus Gidéh.

It stopped raining. The sun came down ferociously. The wet sands of the yard were feathered with steam.

Kumis walked over to the shelter where he had his fighting cocks. He picked up a cage and brought it back. He took his favourite rooster out of the basket, that wiry little bronze bird. Its scarlet comb stood up belligerently.

"Good," said Kumis. "Very good."

Ida Bagus Gidéh rolled up his green cloth, tucked it under his arm. With a gracious gesture, he took his leave. What a fine looking fellow he was as he walked off across the yard, the sun glistening on his shoulders. What a fine-looking couple they would make some day, he and one of these pretty village girls. A pretty girl like Renang? Ida Bagus Gidéh could marry anyone he wished. If a low caste man should marry one of his sisters, that bounder would be banished from the island. But Gidéh could marry a Sudra if he wanted to.

Kumis looked fondly at his bronze rooster and put it back in the basket. He walked over to his little house and came back with a leather packet, unrolling it. In it were neatly packed six blades, like a surgeon's mysterious tools. One had a wavy, serpentine blade, like a *kris*. They were the longest gaffs I had ever seen. Up in Manila they were only three inches. I picked one up, long and narrow, sharp on two edges.

"*Tida brani,*" said Kumis. "Don't dare. He would cut his own throat."

Kumis picked one out. It was slightly curved, sharp as a needle, with one edge like a razor. It was very slender, and five inches long. Kumis felt the gaff's edge with his thumb.

"*Baik,*" he said. "Good."

He put a leather sheath on it, and stowed the others away in his belt. He picked up the rooster's cage, and we walked down the road.

We turned into a narrow passageway between two walls, and walked a long way beneath great overhanging banana leaves. We came into a secluded courtyard, where there were many bell-shaped baskets. There was a *balé* there, and in it men were standing in a circle. Between their legs I saw the flash of feathers.

The ring opened to make room for us. The birds were going to it in a confused whirr of feathers and flash of gaffs. They stood apart, heads low, watching. The red bird darted forward. The black cock turned, with a squawk, and fled.

"Which finishes that fight," I said. I had learned my cockfighting at Manila. "I bet his owner's sore."

"Wait a minute," said Roosevelt. "We'll see."

The handlers had picked up their birds. Holding them, they set the cocks down close together at the centre of the floor. The man who seemed to be the umpire took a cage and put it over them. All stood back. Now the black bird had to fight.

We couldn't see much through the lattice of the basket. There was a flapping of feathers, a war-like clucking. The basket creaked, and edged across the hard-packed earth. Then all was still.

The umpire lifted the cage. Upon the earth lay the russet cock, and over him stretched a coal-black neck, triumphant.

I turned, and Ida Bagus Gidéh was standing there. In his arms was a great white rooster, and over his shoulder swept its long tail, like a defiant plume. His wings were flecked with black, and from his jowls hung blood-red wattles.

A new fight was starting, but we had other things to do. Kumis had gotten out his bird and let me hold him. He struggled a bit in my grasp, and beneath the feathers I could feel his high-pitched muscles. He twisted his head around to look at me.

Kumis took the rooster's left leg. He had a spool of scarlet thread. With many turns Kumis bound on the war-like gaff, tight to the rooster's natural spur.

The floor was clear. The fight was ready. Kumis dug into the rolled top of his *kain* and brought forth a handful of silver *ringits*. He dropped them on the dirt before him. Ten dollars, twenty-five guilders, a month's wages. Ida Bagus Gidéh did the same.

"I feel like a piker," I said, "but I've got ten guilders for Kumis's bird."

"Taken," said Roosevelt, looking at the stout legs of the great white cock.

Kumis and Ida Bagus Gidéh stood within the squatting ring of watchers. They held the roosters by the bodies out before them. They swung them close together, arousing their wrath. The ruffs flew up. The bronze bird's ruff was iridescent with green and crimson. The white cock looked like Sir Walter Raleigh in a rage.

Kumis and Gidéh squatted and placed their roosters on the floor. They set them down a yard apart, not quite facing each other. Their beady eyes gleamed out from the sides of their heads. They could see better on the quarter. Kumis and Gidéh stood back.

The bronze bird looked around unconcerned. His ruff settled, he took a step or two. All was peaceful. But suddenly he wheeled about. The roosters faced each other, heads down, ruffs aflare, tail feathers high. They stood like rock.

All at once there was a whirl of feathers, a flash of gaffs. It was too fast for the untaught eye. Yet there was something indescribably lovely in this moment's clash. Deadly beauty. A sweeping wing, a curving tail, bronze feathers flying in the air. An instant's concentrate of mortal fury.

They came apart, sparred and feinted. The white bird was taking the offensive. He was the bigger, stronger. He darted in to mix. Now the bronze was in the air with a flash of wings. White feathers flew, down drifted in the air. The bronze wings wheeled in flight, landed in reverse, eye to eye with the foe.

The white closed in with kicking feet. Bronze whirred in retreat, and stood a yard away. He staggered. His right leg trailed uselessly. It was broken. The white bore down upon him for the kill.

I felt in my pocket for my money.

The white cock bearing down upon the bronze. The bronze standing his ground, beating wings, staggering away to safety. His foot dragging, bent backwards at the joint.

Why didn't Kumis give up? Stop it, Kumis! This is the time for the white feather if ever there was one.

Five minutes it went on, a game bird fighting for his life. The white bird struck, and the bronze bird whirled his wings away. The white bird struck. The little fellow beaked him by the comb, hung like a terrier in a flapping scuffle.

We were all on our feet now, standing back. Give those gaffs plenty of room, for safety's sake.

The white bird shook him off. The white closed in. Low over his head flew the bronze, with a flash of green and crimson. Lucky escape! It wouldn't be long now.

The white stood like a marble image of menace, as if waiting for the kill. The bronze stood waiting. The white stood there and did not move. The white was drooping. He fell over dead.

Ida Bagus Gidéh picked him up. A tiny red spot showed upon his back. His wounded foe's gaff had cut him neatly to the heart.

Kumis picked up his money. Carefully he took up his rooster in his arms. Kumis held him tenderly and examined his hurt. The joint hung by a shred. The gaff had hacked it almost through.

A fine bird, this game cock. He never seemed so fine and game until today. He turned his head up towards Kumis, clucking. Kumis did not look very happy to win his money.

Kumis unbound the gaff from the cock's good foot. Kumis held the game bird's head and mercifully pricked his neck. His life dripped out.

Some women of the village had gathered, watching us. Madé was there. Kumis gave her the chicken. It was just a chicken now, and probably tough. There's one thing about this kind of gambling. A man may lose everything, but he still has his dinner.

I saw Ida Bagus Gidéh talking to two women, a dumpy one and a tall scrawny one. He gave the fat one his bird, and she idly began plucking it. The scrawny woman looked as though she had a temper, and she was talking very fast. Ida Bagus Gidéh walked away with them.

"Who are those women?" I asked Kumis.

"Those are Ida Bagus Gidéh's wives," said Kumis. He pointed to a handsome little boy. "That is Ida Bagus Gidéh's son. He has two boys in school."

The day had gone dark for me. I watched Ida Bagus Gidéh walking away with his scrawny wife and his dumpy wife. Life is a bungler. Fate is a sloppy artisan. There was Ida Bagus Gidéh, gay and well-favoured. He had his pick, and . . .

I knew a man once who lived in a wilderness. He had a wife who was the only white woman for many miles around. She was the only white woman for miles around, but she had thick ankles. My friend would take me out on his porch and sweep his hand at the jagged horizon. "Isn't it a gorgeous view," he said. He had a gorgeous view, but his wife had thick ankles.

"What made me think of my friend in the wilderness?" I said to myself as I watched Gidéh depart. "He had one wife with thick ankles, but he didn't have two."

IX

"BALI was not always an island," said Kumis, one night when he was more talkative than usual. "Once Bali and Java were all one."

Long ago, even before Majapahit (went Kumis's tale) there lived in Java a Parekan who was the most valued noble in all the court of his Raja. In all matters of tact and diplomacy he was supreme, and he was wise in his judgments of the law. Brave in war and skilled in stratagems also, he led the army of the Raja in great victories. So he became great in that Raj, and master of many *sawahs*. For land was his passion, and he never could get enough of it.

There came a time when the Parekan won a great victory and captured many towns from the Raja of Solo. The Raja pondered long, on what would be a fitting reward. For the Parekan was very powerful, and the Raja was loath to give him yet more land.

Then the Raja thought of his horse, which had been sent to him by the Emperor of Siam. The like of it had never been seen in Java. It was a great beast, tall as a man, towering over all the other horses. Its hair was white and hung long in mane and fetlocks. Caparisoned in gold, it made a proud and kingly mount. But the Raja was getting old, and no longer fit for riding.

"O Parekan," said the Raja. "I have thought long over what reward to give you. I thought of land, but you have so much of that, that more would mean nothing."

The Parekan's heart sank, for more than anything he wished for land.

"I thought of my young daughter," went on the Raja, "but already you have two of my daughters.

"This gift must be princely above all things, for you are a great conqueror; and it must be personal, for you are my friend. So I have decided, good Parekan, to give you my great white horse. It is a king among horses. And when you ride, you will stand out above all others, that men may see how great a man is the Parekan."

Whereupon the Parekan was so overjoyed that he almost forgot his disappointment, for now he would ride even higher than the Raja. The horse was led before him, and he leaped astride. The horse pranced and wheeled, and away rode the Parekan, in delight of his new possession.

He galloped down the roads, and leaped over ditches, and pranced along village streets. The people stared in wonder at the great Parekan and his regal mount. He galloped down through the Raja's rice fields toward the sea.

Then, in leaping a ditch, the great horse stumbled. Headlong went the Parekan into the mud of a paddy field.

And when he arose he found the horse's neck was broken. The horse was already dead.

The Parekan squatted down beside his horse and mourned.

That evening the Raja's men came to him with a sad tale. The Parekan all day, under the hot sun, had squatted in the rice field beside his horse. He had refused food and drink. He was still there in the night. He refused to allow anyone to touch the body of the horse, and the longer the horse remained there in the heat the less anyone desired to touch it.

The next morning the Raja sent messengers to the Parekan, summoning him. But the Parekan would not come. The Raja, fearing for the welfare and the sanity of his valued noble, made the journey himself that evening. There, beside the body of the horse, squatted the mourning Parekan. The Raja stood afar off, for the horse had lain through two days now, beneath the torrid sun.

"O Parekan, why do you stay here?" he cried in a loud voice. "The horse is dead. You cannot return the life to him."

"O Raja, I have proved unworthy of the gift you gave me. Three days I must grieve beside the body of this wondrous horse."

The Raja pleaded with him to come away, but could not prevail. Finally the Parekan said:

"This land where the horse lies must be holy, and I must bury him and raise a tomb in memory of your princely gift."

The Raja agreed.

"And," continued the Parekan, "it must be a great tomb. A temple, I shall raise for it. For that I must have land, extending even as far as the stench to which you so object."

Again the Raja agreed. In the morning he would send three good and honest men, with unimpeachable sense of smell, to stake off the land.

"You shall have this land for whatever you will," said the Raja, "only come now away from here."

"No," said the Parekan, "I shall stay the night."

In the morning the three honest men came, and with them the Parekan set out toward the south. A mile they walked, and the three men marvelled, for they still could sense the vapours of the horse. And on they walked for many miles, until they climbed a long steep hill. The Parekan, weak from lack of food, lagged behind. And suddenly the three men realized that they no longer smelt the horse. They turned back to the Parekan.

"Tuan Parekan," they said, "we have come to the end of this marvellous scent. Here we smell the horse, but yonder we did not."

So they drove a stake and turned westwards; and the Parekan agreed, reluctantly, for there were fine rice fields just a little farther south.

Again they walked for many miles, while the Parekan struggled manfully to keep up the pace. At last they grew worried, they had come so far. The wind must have changed. They had better turn north. Reluctantly again the Parekan agreed.

It was almost nightfall when they came again to the

sea. The tired Parekan looked joyously over the many miles of *sawahs* that were his. As far as he could see were his rice fields and coco groves, and also there were many towns. And while they stood there the Raja, puzzled by the long delay, came to them. He marvelled when the three good men told him of the day's miracle, then thanked them and sent them away.

"Your shrewdness has often served me well," said the Raja, turning to the Parekan with a smile. "Is your wit now turned against me? Well, I have agreed. Take the land. But ask me for no more, good Parekan."

Whereupon the Parekan unrolled the top of his *kain,* and took therefrom a piece of horseflesh, and threw it very far away.

So the Parekan prospered, and was wealthy with the produce of his lands, and you would have thought that a man so wise and successful would have been very happy. But he was not. He had a great misfortune in his life.

All his wives had borne him daughters, but only one had given him a son. This son, the darling of his father's life, had turned out to be a weakling and a wastrel. And what was more, he would not marry.

All the most beautiful virgins that he could find the Parekan offered to his son, but the young man would not look at them. The most skilled and famous courtesans also he sent to tempt the young man, but they were turned away. The old man was sorrowful, for there would be no heir.

Yet he gave the young man everything he could want. He gave his son a palace in the midst of wide miles of *sawahs,* and the young man gathered about him a great band of young wasters who spent his money lavishly. Music was played there and there was dancing, and all day the young men disported themselves with gaming and cockfighting. But it was a house without women. There were boys to do the work of maids.

Vast sums of money the Parekan poured out to pay his son's expenses and his gambling debts. He had to pay nearly all the income from his vast estates. But one thing he would not give his son. That was land.

The Parekan begged his son to marry, even if only to get an heir.

"Give me land if you wish me to do this," said the son.

But the Parekan would not.

So it went, and it became more and more a strain to the Parekan to pay the debts of his son, and of all that household of wasters. At last came the great day of the racing of the belled bulls. The Parekan's son had a black bull that was greater than any other, with long legs and iron thews. He depended on it to win for him, and pay back all his losses. He wagered heavily.

All the country-side was there to see the racing of the bulls, as they dashed across the dry, yellow *sawahs* that had been levelled for the course. The bulls rushed madly, and the great wooden bells that hung from their necks rang out in deep-throated music. The black bull was leading, winning easily, when suddenly he stumbled and fell, just as the Parekan's horse had fallen.

An Arab came next day to see the Parekan. The son had been getting deeper and deeper into debt, said the

Arab. Here was a loan from three years ago, that had doubled with the interest. And here was another of five years standing.

"These debts are so much and so much," said the Arab. "But yesterday your son made still larger wagers, and gave vast odds. And for security he pledged the lands about his house. Now that he has lost, we find that the lands are not his, as he said, but are yours. You must pay these debts and these wagers."

"But to do that I should have to sell my lands," cried the Parekan.

"Then you must do so, for it is on the honour of your house."

The Parekan was panic-striken. He rushed to the Raja, but the old Raja laughed, and told him now he could sell his land. The Parekan went back to the Arab and told him he would pay in thirty days. Then he went to his house, put off his fine garments, put on a simple *kain*. With only a little bronze bell and a bag of incense he set out walking eastward.

Four days he walked, a lean old pilgrim, ringing his little bell; and at night he slept in the villages by the way. On the fifth day he came to a place where a narrow strip of sand ran out between the waters. Hungrily the waters bit at the sand, but could not eat it. The Parekan walked across, and beyond was Bali.

"Where are you going, pilgrim with the bell," asked a Padanda beside the road.

"I go to visit the great Naga that breathes fire in Gunung Batur," said the Parekan.

"That is a great Naga in the crater of Batur," said the Padanda. "The only one greater is that Naga which encircles the earth and which sometimes swallows the eye of day and the queen of night. But unless your petition is very difficult, you don't need to go so far as Batur. It is not so far to the mountain that split when the gods laughed, and there lives another Naga."

"Oh," said the Parekan, "I had not heard of that Naga. I will try him first."

So he went to Gunung Kawi, and burned incense where the hill was split. The Naga was pleased, and came forth, looking as amiable as it is possible for a dragon-serpent to look. The Naga rolled his vast eyes, and showed his great fangs, and all his scales glittered in the sunlight, with gold and precious stones.

"O shrewdest of Parekans, who once rode the great white horse," said the Naga, "why do you come thus afoot and in poor raiment?"

"Naga, I have come the point where my shrewdness is of no avail, and I am helpless before the cruelty of the world. I need the help of the great Naga, who is wiser than I, and indeed is the wisest of Nagas."

"And what do you want?"

"Naga, I come to ask advice."

"Oh, tell me what the trouble is." And the Parekan told all the troubles that his son had caused him and all about his son's strange behaviour, and how he was about to lose his land.

The Naga was pleased that the wisest man of Java had come to him for counsel, and for a long time he talked advice. The Parekan had made great mistakes in bringing up his son, he said. He had favoured him too much and pampered him, given him everything he wanted instead of letting him grow up among the children around the place and learn his proper lessons. It was a great mistake. said the Naga, to have whipped the boy that time he got petulant. No wonder he grew up queer. But the way to treat him now was to refuse him any money at all unless he married. As the Naga talked he got more and more interested in the case, and he talked for a long time.

"Thank you, Naga," said the old Parekan. "Your words are a great help. I know just what to do. But there is just one little point we have overlooked. How am I to pay these debts without selling my land?"

"Well," said the Naga, shrugging his shoulders. "I had overlooked that little point, hadn't I? Surely, we can't let our plans fail because of a detail like that."

Shrugging his shoulders again, the Naga shook himself, and down from his back came a shower of gold, diamonds, and rubies. Picking up his *kain* like an apron, the Parekan had soon gathered up a great weight of treasure. When he had found it all, he turned and said:

"Naga, I can tell you truthfully: as an expert in diplomatic missions and in the wisdom of state, I have never in my life known any advice so valuable as yours."

And the Parekan returned to Bali.

He paid off the debts, and he sought to follow the Naga's advice, and refused money until his son should marry. But the son did not take kindly to this.

"You needn't tell me there isn't more treasure where that came from," said the son.

"That was given me by the Naga of Gunung Kawi," said the father, "and he will not give me any more. He said I was not to give any to you."

"Well," said the son to himself, "if the old man can get treasure there, I expect I can get it too."

So the son arrayed himself in his finest raiment and set out for Gunung Kawi. He took three of his men campanions with him. He took a small boy also, to carry his *kris*. He crossed over the spit of sand and came to Bali.

"Naga!" he cried, when he had reached the serpent's place; and the Naga came forth, snorting at this impudence.

"I am the Parekan's son and I have come for treasure!"

"Upstart!" cried the Naga, and in one whirl had entered his hill again. But his tail remained out, and there on its end gleamed a vast ruby such as the Parekan's son had never seen. Angrily and covetously he snatched his *kris,* and with one sweeping blow he cut the ruby off.

But he no sooner had picked it up than the serpent had whirled again, and with one nostril's blast of fire the Parekan's son was no more than a dark greasy spot against the rock.

The Parekan, on hearing this sad news, set out again for Bali.

"Naga," he said, "I have followed your advice, and see what sadness has come to me."

"I sympathize with you," said the Naga. "I will give

your son back. But it must be agreed that he never again return to Java."

The Parekan agreed, and with one inhalation of the Naga's breath, the worthless son stood before them.

"Now, see what trouble your stinginess has got me into!" cried the young man when he saw his father.

"The Naga has given back your life," said the father patiently, "but he says you must always remain in Bali."

"What!" exclaimed the son. "Stay here, away from all my friends!"

"Oh," said the Naga, yawning. "We have polygamy in Bali too; so I fear, though I am sorry to say it, that since there are not enough women to go around twice, you will find companions here."

At that the Parekan embraced his son in farewell, and set out for Java. He came to the long, narrow spit of sand, and as he walked across, the waters lapped hungrily at right and left. The Parekan stopped. Before him lay Java, and all his *sawahs*. Behind lay Bali and the son who must stay there. Would he be content to stay?

The Parekan looked at the soft sand beneath his feet. He extended the little finger of his left hand. With its long, long nail he touched the water's edge; then across the sand he gouged a slender line. The waters rushed in, and mated hungrily. The Parekan walked on toward Java.

Behind him a miracle had happened. Ravenously the waters fell upon the sand, and soon there was no sand left at all, only the dark, surging waters that are the Straits of Bali.

But the Parekan was not watching this. His eyes were straight ahead. Far away before him were his great terraced hills, trickling with water and sprouting green with rice.

X

THE story tells that many years ago a Jesuit missionary came to Bali. He was an amiable soul, and the people loved him and were his friends. When he learned their language he told them about his God. They were very much interested; and then they told him about their own Holy Trinity and the system of religion from which sprang all their glamorous culture. They were interested in his tales about the saints; and they recounted the stories of their own Arjuna and Samarna, and other men whose greatness and goodness had elevated them to places among the gods. Somehow he made no converts.

After two years, however, a high-born young man named Kramas espoused the Christian faith. The good father, elated by his success, wrote home about it; and his superiors, thinking it an auspicious moment to bring this people into the fold, sent him an assistant.

This new soldier of the Cross was as vigorous as the first was genial. His eye was fiery, and his words, and he told of a fiery hell. The natives listened, but were not alarmed. They knew about the fiery hell already, had painted excellent pictures of it with punishments of piquancy quite beyond a Jesuit's imagination. They listened, but their smiles gave way to blank expressions of bored tolerance. He drove upon them with a fierce will, and all the arts of evangelism, without avail. So he called Kramas, and told the young man that he must bring other converts. It was not sufficient to believe; one must spread the Kingdom of Heaven to be saved. Kramas tried and failed, and the more he failed the more the new priest urged.

Kramas was bewildered. He had left his faith for a God of love and mercy, and now he was condemned because others would not follow him. His people were cold to him, for he had deserted his gods. And still the blunt priest urged. The youth was troubled, and a fierce resentment swept through him. In his hysteria he went to the house of the priests at night, and with his *kris* he killed them.

Now he was even more troubled, and a cold sanity came over him. He had killed two men; and was it not the law of Vishnu that a man should do no murder? After death he would face the fierce-eyed Durga, with her hand upraised, and then Yama, with the wheel of law. And he would be cast into outer darkness, and never more would his immortal soul know the heavenly realm of Dhanapoti, king of wealth. For his deed was written in the thirteen books of the gods. But still more was he troubled, for he thought of his people. Two white *tuans* had been killed. The Dutch would blame the nation and their vengeance would be swift. So he went to the temple, where he had not been for a long time, and prayed. Then he went to the Pungawa and said:

"I have murdered those two *tuans* who preached their God. Let me then be killed according to the law, that the *Orang Blanda* may know we are a just people, and not seek vengeance."

So the men of the Pungawa drew lots, and one of their number was named to execute Kramas. On the appointed day he dressed in his finest *kains,* draped about his waist and under his armpits, and at his back was a bright-handled *kris*. Kramas came before him. The men of the village escorted the murderer, but he wore no fetters, and was not bound in any way. He stood before the executioner, who said:

"You have done me no harm. But you have committed murder and the law says you must die. I have been appointed to kill you. Am I permitted to do so?"

"Nguh," said Kramas, lifting his chin and eyebrows in token of assent. "Yes, you are permitted to kill me." The *kris* was raised, and struck him through the heart.

I know that Kramas died in this manner, for this was the way all condemned criminals died when seven Rajas ruled Bali.

After that no one tried to make Bali Christian.

Kumis knew nothing of the modern grandeur and philosophies of Christendom. Though he was the slave of nature and his domestic arrangements were primitively close to the animals, it never occurred to him to doubt he had a soul. His soul had come to him when he was a child, and even in maturity it was likely to wander from his body while he slept, and perhaps enter into the body of some animal for the sleeping hours. Kumis would never kill a chicken after sundown. It might be the lodging of a sleeping friend. Kumis's soul was immortal. It had lived in many past generations, and would live in many generations of the future, always as a human being. Kumis was careful of his soul, for his deeds were written in the thirteen books of Heaven. If he were evil, he might be cast into a dark place and perhaps never again return to Bali. Kumis hoped to go to one of many Heavens, according to his merit, then in later years return to Bali, a child in his own village, his own family. Kumis loved Bali, and though you might offer to pay him well, he would not leave it.

As I walked out to bathe, on my first morning in Kumis's home, a flurry of wings suddenly flapped across my face. I stopped short, startled. Then I noticed I was standing beside a small plain wooden shrine. The little box at the top opened toward me, and in it was a setting of hen's eggs.

I went on toward the well, observing that religion might be practical.

A few days later, after the old hen had taken to brooding her chicks around the yard, I saw something else.

All the family was kneeling in a semi-circle about the shrine. Old Sintang, mother-in-law of Kumis's sister, was sick. The priest's sister was there, in a white *kain*, and there were flowers in the knot of her grey hair.

Madé, Kumis, and his sister, Nur, held out their hands and the priestess put frangipani petals in them. She sprinkled them with holy water and put water in their hands. They touched the water to their lips. They touched their fingers tip to tip and bowed their heads. Nur took an offering of palm leaves, all cut in fanciful designs, and put it in the shrine.

The next day Sintang was up and around again.

This little shrine was the domicile of an unimportant deity, the *déwa ruma,* god of the house. Gods of more dignity inhabited the temple adjoining Kumis's house, the village temple, the temple of the dead, the temple at the priest's house, and all the other temples that clustered around the town.

In India is the Trinity—Brahma, Vishnu, and Shiva. In the the north of India Shiva is dominant, and in Bali also. In India, Shiva the Destroyer. But in Bali ferocity is gone. Surely here, in this amiable island, Shiva is the Preserver, who has kept this civilization from the Crescent and the Cross.

I asked Ida Bagus Gidéh about his gods.

"My father has given me sacred books to read, written long ago in palm leaves," said Ida Bagus Gidéh. "They end with the words: 'You who have read these things, speak no word of them to any man.' It were better that you speak to my father.

"He is in Badung this morning sitting as Magistrate in the court, but when the eye of day is about there I will take you to him."

Ida Bagus Gidéh had lifted his hand, was pointing to the sky of afternoon.

For centuries the vedas, sacred formulae of Hinduism,

muttered with strange motions of the hands and sprinkling of holy water, have been the secret of the Brahmana, the charm by which these priests have held their pre-eminence of caste. In Bali, as well as India, the priests still mumble their vedas, archaic words and sentences whose meaning is lost in antiquity and in the confusion of India's gods with the ancient nature gods of Bali.

That afternoon Ida Bagus Gidéh led us to his father's house. The Padanda Ratkarta lived in the *banjar* of Kedaton. He was priest of half a dozen *banjars* in the neighbourhood, but not of Kedaton. When they wanted a priest they had to send elsewhere. Ida Bagus Gidéh had nothing to do with the people of Kedaton. Being a connoisseur of religious quarrels I tried to find out what this was all about, but never could do so. Ida Bagus Gidéh never understood when we asked him about it.

We came to what had been a tall pillared gate, whose crumbling bricks had fallen all apart. By each pillar stood a great demoniac figure, whose fierceness had been softened by moss and years' erosion. Inside was a tree shaded courtyard, whose houses were mostly equipped with walls, and we passed through into a temple sweet with frangipani trees.

There, in a *balé*, sat the Padanda. A chubby man, whose long grey hair was in a knot on top his head. His stomach pouched out over his white *kain*. He was a genial looking old fellow, with a scraggly grey moustache. He was drawing something on a board.

When we had been presented he showed us what he was drawing. Here was a snub-nosed, long-fanged demon with

arm outstretched in the attitude of killing, and before
him crouched a peak-nosed prince. It was the conventional
style of Hindu mythology, seen everywhere in Bali and
Java, figures much like the puppets of the shadow plays.
They were there in that *balé* yonder, on that sacred
painted panel of red, blue and gold. These figures in the
Padanda's drawing were conventional, but the composition and draughtsmanship were the priest's own. This
was fine craftsmanship, and it was but the casual pastime
of a pot-bellied village parson.

The Padanda was pleased at our interest. He reached
into his belt and brought forth a roll of paper.

"I am building a new gate," the Padanda said. "The
earthquake ate the old one. The men are making bricks
now."

He unrolled the plan he had made for the new gateway.
It was just a rough pencil sketch, of two pillars with
gentle curves, and a faint hint of the carving that would
be upon them. And from this sketch the artisans of the
town would make his gate. Probably as they worked he
would tell them to do it this way or that way. I knew the
gate would be finely sculptured and finely proportioned.

There were shrines opposite where we were sitting.
One was a wooden house on stilts, with closed doors
painted blue. I knew there was nothing inside those doors.
This was a place for the god to dwell. There was a high
stone seat also, soft with moss, with many carvings. No
one ever sat in this chair. It was a seat for the god.

The walls of this enclosure were studded with the upright stones of Shiva's lingam, the eternal phallus. I found

nothing to shock me in this, more than in Methodist church steeples and the Washington monument. I knew of the Druids. But I was reminded of strange carvings I had seen on temple walls, weird ludicrous obscenities.

What about them, Padanda Ratkarta? What do they mean?

"Those, *tuan,* are the bad doings of Raxasa. She is a very evil spirit."

Evil spirit, evil spirit. Why so many grotesque figures, so much talk of evil spirits in this land where the unpleasant moods of nature are at an almost irreducible minimum. Why do you believe in evil spirits, Padanda?

"If I go into a dark and lonely place, *tuan,* I am afraid.

"Something must be there, or I would not be afraid. It must be evil, or I would not be afraid.

"I cannot see it, so it must be a spirit."

I asked the Padanda about his gods.

"There are Brahma, Vishnu and Iswara," he said, "and these are the same as Brahma Shiva, Sada Shiva, and Prama Shiva, and all together they are Tintiya."

The priest pointed to a figure carved on the base of the great stone throne.

Tintiya! Sanghyang Tunggal! God of many names. This was his image. Weird, spindle-limbed, round faced creature, with his foot upon a wheel, and bursting from him everywhere were triple tongues of fire. Primitive, terrifying glimpse of man's childhood, with a childish laughter about him too. Tintiya, God above all Gods. The all-seeing, all-living Presence. The all-permeant ether, Aakaasya of Sanskrit. Tintiya, the one God whose only task was to exist. Was this primitive, tribal image become man's vision of monotheism?

Who is Tintiya? I had asked many men. Tuan Allah, they replied. Confusing, this. A white man had told me that the Balinese were so tolerant that they sometimes carried offerings to Tuan Allah, the Mohammedan's God.

"Who is Tintiya?" I asked Ratkarta. "Tuan Allah," he replied. But, I continued, we were talking in Malay, a Mohammedan tongue; was he, perhaps, just making a rough translation? Was this Tintiya the Mohammedan God?

Now he understood. He smiled, a patronizing smile, I think, this grey man who knew the vedas and the sacred books. Tintiya, Allah, Jehovah. Could there be more than one God above All? He waved his graceful left hand, with its curling thumb-nail three inches long.

"They are all the same," he said.

XI

THE child of Bengkel plays upon the grass, and from the bamboo leaves and palm trees dewdrops fall upon his brow. Through this baptism a soul enters into his naked little body, gradually, day by day. It may be the soul of his grandfather, or of an uncle who died many years ago. This child is treated gently. If you should whip him you would drive his soul away, and then he would become a terrible thing.

There were always a good many children around Kumis's house, his sister's and the neighbours'. Babad had plenty of playmates. What attractive children they were. I should like to have a Balinese baby, though I have no hope for one.

It was hard to get acquainted with them, though. They were so reserved. I wanted to dandle them on my knee, but never did. Sometimes I forgot myself and threw a friendly arm around a little boy or girl. I felt as though I had tickled a Bishop's wife. A Methodist Bishop's.

These children were always happy and gay, in a calm manner. Their smiling features had a certain delicacy that they would partly lose when they grew older, and there was an innate gentility about them. Nobody ever had spanked them, and they were never naughty. They never quarrelled among themselves, or got on older people's nerves. Only once did I hear a baby cry. That child was ill, with a carbuncle on his arm. I wanted to take him to *Tuan Doktor,* but they wouldn't let me. I expect he got well.

Babad and his friends carried slingshots salvaged from rubber tires. Sometimes they would fly kites, and fight with them, tangling strings, to see which could remain in the air the longest.

Kumis's little niece Nioman Renkog would carry her naked baby sister astride her hip. The babies never wore anything but bracelets. Some of Babad's playmates carried their little brothers on their hips. Some of these boys who carried their brothers were almost as old as Babad, and they hadn't started to wear anything but bracelets. How they flourished!

Other wonders I have told, and been quite generally believed. People are trustful about matters of which they know nothing. But about this matter of children I find scepticism. People know children. They know that babies cry, that older children fight and rebel against authority. At times I have found myself almost doubting my own memory. This child life of Bali is incredible.

Yet lately, in reading some modern views of child psychology, this mystery has seemed less obscure. No child of Bengkel ever has a chance to be pampered. No child has more than a minimum of adult attention. No sooner is an infant weaned, than it goes into the care of another child. It finds itself immediately part of a child society, to which it must adjust itself. And thus from the first dawning of intelligence it develops a completely social prototype. The notion that whipping drives away a child's soul would seem to be merely another way of stating a modern view of educators.

This early development seems to be the root of that perfect social adjustment which underlies the peculiar peace of Bali. But it is not a complete key to the mystery, for relatively similar circumstances surround the children of more turbulent primitives. I have spoken elsewhere of heredity being at the root of racial differences in character. But as the child of Bali grows older, he grows into a social inheritance which may be even more important than any peculiarities of his chromosomes.

Isolated in the mountains is a village which is not Hindu, which in all probability is a fossilized relic of aboriginal Balinese society, an uncorrupted epitome of national character. There must have been a stern conflict of ideals in those ancient centuries when the Hindu hordes came to Bali, before mutual assimilation produced the modern culture of the lowlands. In this isolated mountain community the 100 per centers won. The people live by their ancient handicrafts, unchanged by modern methods. Their vegetable dyes hold their warm rust red until the fabrics fall to pieces. By penalty of exile they keep clean from all outland immoralities and divorce. By exile and deportation they keep themselves purged of all radical economic heresies such as the private ownership of property. Their society is a simple, unadulterated communism, and such is their social adjustment in the group that they live quite naturally in that state of co-operative solidarity which Christians and Russians so futilely preach to turbulent white men.

Just this inheritance lies at the foundation of Balinese society. Over it is the caste system, an alien and almost meaningless excrescence. Private property is an established institution. But underneath is the very real identity of the group whose life centres around the *balé banjar,* which harvests the rice, builds the temple, plays music. Kumis and the others in Bengkel exist less as individuals than as threads in the social fabric of their *banjar.*

The Dutch had some difficulty before they recognized this. A *controleur* in North Bali many years ago tried to enforce certain rules which conflicted with the inviolable customs of the people. That complete acquiescence to authority which is part of the Balinese social adjustment was, in this case, not strong enough to break their adamant conservatism. They did not obey, and a few of them were arrested. The next day before the jail at Singaraja appeared a large crowd of men, some two hundred, I believe. They were neighbours of the prisoners and they also had violated the law. Equally guilty, they insisted upon being arrested. There was no room for them in the jail, but try as they would the Dutch could not drive them away. So they were arrested, and the Dutch finally broke their spirit by shipping them to Java: for no Balinese can bear to be away from his own island, his own temple.

But from the incident the Dutch learned not to be too insistent in their efforts at reform.

Thus (except for the special case of royalty, which always has been somewhat turbulent and warlike) the Balinese child grows from his infantile play group into a perfectly adjusted adult society.

But I am not such a fool as to make any dogmatic analysis of the psychology of Bengkel. Between the brown man and the white there is a gulf too deep to fathom. And he who would enter into the eastern mind sufficiently to understand it fully, must perforce discard his western identity and thereby lose his power of translation. Even in a superficial glance one encounters psychological difficulties. Granted that the child of Bengkel develops a social prototype that would satisfy even Alfred Adler, what then is his "goal of superiority"? I never saw any indications of struggle against individual feeling of inferiority, unless the vicarious conflict of cockfighting and the gorgeous display at cremations be considered as such. Rivalries between villages there were, but I never noticed them between individuals. Individual striving in the arts seemed quite divorced from personal aggrandizement, and was done anonymously with little thought for preservation of its accomplishment. Doubtless the individual craftsman gained standing among his fellows, but he was not working for posterity. Is it possible, I wonder, that a religion which guarantees immortality and foretells a mounting through seven heavens may satisfy an individual's personal aspirations?

But these are afterthoughts. Such theorizings did not occur to me in Bengkel, and indeed they would have seemed out of place. Certainly Kumis had no use for such analysis. He knew nothing and cared less about Vienna and its philosophies.

At any rate, I pitied Kumis, though I never told him so, that none of these children around his house was his very own, that he had never been able to give a feast in thanks to the gods for birth. But one day I learned he might have had a worse misfortune.

Kumis came with a sad look on his face. A great calamity had befallen a village across the rice fields. Two babies had been born at once.

"But Kumis, you rejoice when one baby is born. Isn't it a much greater blessing if there are two?"

"Oh yes, if they were two girls that would be great feasting, and if they were two boys that would be even finer, and in the old days they would have been given to the Raja. But these born yesterday were a boy and a girl, and that is a very evil sign."

So the twins' family was in disgrace, and their village was in fear of the gods' displeasure. The house where they were born must be torn down and burned. Their mother must take them away and live outside the village for six weeks, until she and they were purified.

For it was not right that brother and sister should have lived so intimately.

XII

THERE was scarcely anything about Runis and Gusti Madé Réi to distinguish them from the other little girls who played about the village and came to call at our *balé*. You might have noticed Madé Réi because she was so beautiful and had such laughing eyes, or Runis because she was such an ugly little thing, with such a serious mien. Madé Réi was a Gusti, of the Wisia caste, but nothing except her greater refinement of feature would indicate that she was of higher birth than the little Sudra Runis. They dressed just like other little girls, in a simple *kain* with a white scarf twined through the hair. You would not guess that they were the temple dancers of Bengkel, and of all the country round.

Runis and Madé Réi were nine years old. Because they danced the sacred, traditional measures of the *legong,* and because the men of Bengkel played music for their dancing, the banjar of Bengkel was exempted from taxation.

When they were four years old, they began with other little girls their age, to learn to dance. The older girls of the village were their teachers, showed them how to make strange postures with their hands and arms, to move easily on their heels, with knees bent and toes pointed out. Their hands were naturally slender and supple, with fingers curving backward at the knuckles, and knuckles folding together easily to pass through the smallest bracelet. These gifts were practised, until the fingers might be bent back to a sharp angle with the back of the hand, until each finger might be moved independently of all the others. These and many other things the little girls were taught, for two years, until it was apparent that Runis and Madé Réi were the best of all.

Then the little pair went to Kesiman, a mile away, where there was an old man learned in the dance, who in days of regal splendour had been at the court of the Raja of Kesiman. Three times a week they went to him, and learned all the traditional movements of the *legong*, with its little refinements, just as they had been practised for hundreds of years. Their bodies attained the co-ordination of delicate machines. The dance they learned was much like the traditional dance of Cambodia, rooted in the same ancient culture, but their training was very different. They were not forced to throw their limbs out of joint or assume unnatural contortions like the little girls of Indo-China, nor were their childhoods stolen and imprisoned in a royal court. They continued to live the simple life of their village, and when they turned eight years old they became its official dancers. For Rupag and Renang were now twelve, and too big to continue longer the lightning speed of the *legong*.

Like the other children of the village, these little
dancers were aloof, reserved. It was a long time before
they would treat me as anything but a stranger. Their
indifference was colossal, and I was hurt, for I was rapidly falling in love with them. I was jealous of Roosevelt,
for they would come and lean against his knee and talk
with him.

Sometimes I gave them a little money, after they had
danced. They accepted it in a matter of fact, impersonal
way, as a tribute to their art. They were used to receiving
presents. They never thought of thanking me for it. No
native of Bali ever says ' thank you" unless he has been
exposed to the ways of aliens. The nation's manners are
entirely lacking in those little matters of form, such as
"please" and "excuse me," which are taught to our children as habits, things to be said whether they are meant
or not; manners there consist in an obliging consideration
for others' comfort. If I give a Balinese something, it is
either because I owe it or because I want to give it, and
it is accepted on that basis without demonstration. If
there is reason for gratitude, I may assume that it exists.
A great deal of hypocrisy is thus eliminated. At any rate,
presents would not buy a show of friendship from Runis
and Madé Réi.

It seemed as if I would never be more than a stranger,
when one day a rather idle idea came to me. It occurred
to me, while thinking of the strange postures of the *legong,* that I could strike an attitude even beyond their
miraculous powers. For I was the unfortunate possessor

of an arm left crooked in a childhood accident. Now, scarcely thinking, I threw out my arms to show an elbow weirdly bent.

The effect was electrical. It was an instantaneous hit. I was the toast of the town's childhood. Not an artistic success, perhaps, but then, neither was *Abie's Irish Rose*. For all practical purposes, that elbow was as fertile a source of popularity as Tom Sawyer's sore toe. I had ceased to be a mere strange *tuan;* I had become an interesting person. Around me was a group of laughing children, and the most hearty laughter came from Runis and Gusti Madé Réi.

This popularity was not an unmixed blessing. I was, of course, subjected to a certain pitiless publicity, which was not entirely pleasant. But it meant a barrier torn down, a great gap made narrower. I would be standing in a group, and feel a tickling at my arm. Turning, I would find a pair of laughing eyes, peeking at me from behind a convenient knee. Or perhaps there would be three or four pairs of childish eyes, and it was for me to guess which ones were guilty. Then it was for me to try to catch the daring one, to intercept her light speed with my lumbering. It became a game we played. This was pleasant, though it sometimes palled.

Out of all this was to come, however, the next time the *legong* danced, the greatest thrill I received through all my weeks in Bali.

While she was dancing, Madé Réi *smiled at me!*

It may seem strange to make so much fuss over a dancer's smile. But in this ancient *legong* dance all per-

sonality is submerged. The dancer's face is a mask; and
her eyes, flashing, staring, snapping, are as much a part of
the dance's balanced, formal whole as are her hands and
feet.

The point is this: that in smiling, she was violating all
the rules; and it was *I* she smiled at.

When the girl of Bali prepares to dance in public she
puts on clothes.

The preparation of Runis and Madé Réi for dancing
the *legong* was a complicated matter. First was put on a
long green *kain* trailing on the ground, and all ornamented
with flower patterns in gold leaf. Over the shoulders was
a little vest, and to this were sewed long, tight sleeves.
From breast to hips, like a race-horse's ankle, the body
was bound with a tight overlapping bandage of strong
elastic cloth; and around this again was wrapped another
bandage of gold-leaf pattern. From breast to knees hung
a broad, straight apron. Over the shoulders was a wide
leather collar, elaborately tooled and gilded, and over it
a gold neck-piece, set with coloured stones. The ear-tubes
were of silver, studded with gold. All this costume was
of green and gold. The brow was ornamented with
little discs of white palm leaf, and on the head was a
broad crown, spraying with three pounds of pure gold
flowers.

When all this elaborate costume had been put on them,
Runis and Madé Réi would go to the temple and pray.
Then they were ready to dance.

A great banyan tree hangs over the road at Bengkel, and over the temple gate. Its massive trunk is a solid bundle of small trunks, and its spreading feathery branches are shaggy, festooned with vine-like trailers. If one of these touched the ground, it would take root and grow into a solid, slender new stem, supporting and nourishing the branch from which it hung. Unchecked in its crawling growth the tree would wander indefinitely. Roosevelt discovered one such, tucked away by a remote village in the centre of the island. We paced it off one day, eighty yards one way, seventy the other, larger than the famous banyan of Bombay. We went in through its myriad pillars. There was an open place inside, with a shrine, and the sun struck down through mullioned windows into this natural chapel.

But the trailers of the tree at Bengkel are kept clipped. Their ends are bundled into brooms for sweeping dooryards. The tree spreads out from a bundle of fasces-like pillars, only about fifteen feet in diameter. Beneath it the ground is level and hard-packed.

Bits of sunlight filter through the tree and fall upon the gold-encrusted instruments of the *gamelan*, splashing from the flower-pointed carvings, like the music which eighteen vacant-eyed men now strike trembling from their bronze. The orchestra is grouped behind a golden frame, within which hangs its gong. Flanking the gong, on stools, sit the little dancers, immobile as statuettes.

Close-packed about an open square before them stand the people of the village—women, children, men. There always is a good audience for any public entertainment in

Bali, and admission is never charged. If it is a special occasion the performing group, not its individuals, may be paid, by the village or by some rich man. But usually it is an entirely free affair, a part of communal life.

The hot, heavy afternoon air is burnished with brazen music, a stately repetitious pattern enlivened and varied by silken frills. In the dancing girls' hands are fans. Now suddenly the fans, stirred by an incomprehensible cue, are fluttering. The dancers rise, advance on heels, knees bent and arms outstretched. They weave about the square, their arms make patterns. Patterns, patterns, their dance is a kaleidoscope. What discipline is here! Can these be children, moving so formally? Arm, forearm, fingertip, chin; eyes snapping and glaring; each minute detail is but part of a finished whole. Nothing bursts its bonds. These dancers' youth needs no apology. They are the fine-pitched vehicles of precise and practiced art.

Patterns, patterns, impersonal as a Cézanne sketch. Glorification of lines. There is in this dancing not one jot of sex appeal. Naturally, do you say, since these are children? It would be the same, were they mature. In none of the dances of Bali is there a hint of sex. There is not a hint of personality, and "personality," so psychologists have told me, is but a form of sex attraction. In this is the great gulf between East and West. Have you ever noticed how important is this matter of personal appeal in our western dance? A hoofer comes on the stage and gyrates with astounding perfection of physical agility and poise; his act falls flat. He is followed by one of mediocre technique, whose magnetism stops the show. There is nothing of this in Bali. A dancer may have a touch of individuality, but not one bit of individualism. Runis and Madé Réi are not expressing themselves; they are setting forth a pure design of line and motion. I try to think of a western dancer whose art is similarly unpersonified. I think of Michio Ito, but then remember he is an oriental.

This lack of personal exposition is not only true of dancing. It is especially true of music and art. The Balinese would not know what you meant if you talked about the eternal verities expressed in a symphony, or about interior decoration for personalities. He has not become self-conscious. That is probably why he has no creative literature. The Dutch found great libraries of palm-leaf books when they came to Bali, the Hindu scriptures, legends, commentaries, but in only two cases could they learn the names of writers. Usually the books were copies, and in no case did any image of the scribe shine through his work. We of the West who write are by nature exhibitionists, seeking to show off what is hidden by our unpretentious exteriors. But the Balinese has no such impulse. In all his art there is little of humanism. Time and again it has made me think of Europe's thirteenth century, before the Renaissance, when men were beginning to see the beauty in leaves and flowers, in archways and ascending spires, but had not yet learned from Greece to seek humour and sorrow in the human heart. I like to think of Walt Whitman in Bali. With what barbaric yawping he would have had off his shirt! How he would have snatched these people to his hairy bosom! Children of Adam, gushing, jetting hearts! And how bewildered he would have been, that sublime egotist, when the people of Bali failed to understand him, when he realized that here the individual is but an unaccented beat in a poem that scans as precisely as the most conventional verse.

Runis and Madé Réi are dancing. Their hands are all sudden, unexpected, angles like the branches of the frangipani. Slim, gentle hands. Their bare toes flicker from beneath their trailing *kains*. In this dance is nothing of the light fantastic toe; the feet are earthbound. Perhaps they fly as sixteenth notes beneath their *kain,* but always they are subordinate, mere vehicles of locomotion. The toes are pointed outward, in line with widespread, bended knees. Above the hips is done the dancing. Now I know why Runis's torso is so tightly bound. How it twists and ripples, from waist to shoulders! Unsupported, tendons would be torn. Now I know why these dancers must be tiny. Big girls could not have such winking limbs. See how their fingers, arms and shoulders flash. This is a majestic measure, stately, yet all compounded of a sparkling motion. The flight of the gnat is graceful, slow; but its wings are blurred.

The lambent air reverberates with music and with tropic heat. Two sprites of Majapahit's day are dancing. What was it Einstein said of past and present being all today? These are Arjuna's daughters, ethereal as the flecks of sunlight playing in their crowns. Absurd to think that they are common clay. They have been dancing half an hour and are not half finished. The heat hangs heavy over all the world. What frail childish body, swathed in heavy clothes, could stand this stringent, wearing gait? But now a hand withdraws a moment from the dance. Prosaically it sweeps across a beaded brow. Yes, they are human, after all.

Runis and Madé Réi are dancing, as danced the girls long centuries ago. Once, perhaps, this *legong* had a story. Long it has been forgotten, but I seem to see it now. Two princes are quarrelling. They advance toward each other, breasts stuck out like fighting cocks. They halt, quivering. Their shoulders shake as if in a purified shimmy. They wheel and step apart, with heels kicked out before them. Now Madé Réi is coming back, left hand flexed behind her, right arm straight ahead and fingers pointed. She lunges, fencing. And there now is Runis upon her knees, dying, vanquished. The conqueror proudly marches. But Runis is by the gong, and on her arms is fixing cardboard wings. The music changes, and bass bells sing a thrilling theme. The theme of the Garuda Bird. The prince returns, incarnated as the Himalayan eagle. Runis is squatting on her heels. Her wings flutter. Somehow, squatting, Runis leaps from side to side. Head jerks, shoulders shiver. Shades of St. Vitus, caught and imprisoned in a dance! Triumph to the Garuda, winged justice!

The *gamelan* is silent. Runis and Madé Réi are beside me smiling. They have lifted their crowns to dry their curly hair. They breathe as easily as I.

XIII

ROOSEVELT spoke every now and then about the Old Lady of Kesiman. A beautiful old aristocrat she was, he said. *Anak Agung,* child of the great. Sister of a Raja, she had been once, in days of splendour; but now was all of regal times that remained in Kesiman's half-deserted palace. He wanted to make a picture of her. We were talking of going there one day, when Ida Bagus Gidéh said he would take us.

It seemed in these days as if Ida Bagus Gidéh didn't have much to do but loaf around and be pleasant to us. But this was rainy season. When the dry days came he would be busy, for he was commissioner of the water supply in the district of Kesiman. Daily he would go to Kesiman, administering the ancient water rights, so much irrigation to each terrace. So he was familiar at Kesiman when he took us there.

A brown young man, wearing shoes and a high-collared white coat, sat by the palace gate at Kesiman. He pecked at a typewriter, symbol of tomorrow's doom.

The palace was a great walled compound, with massive gates. Inside there was a lonely feeling. Above the wall beyond I could see the peak of a high pagoda and the tops of frangipani trees. The wall was studded with the phallic cylinders of Shiva, reproduction deified. But over all was a feeling of sterility. There was no Raja now, only the Old Lady.

"Let's go in the temple first," said Roosevelt. "There is a great zig-zag moat in there. It will be dry now, but it is interesting."

We went through a little door in a great closed gate, and found surprise. The moat was full with gleaming water, which cast shimmering reflections of the great pagoda. The *balés* within the moat were crowded with women, who had gathered from all the village. They were cutting patterns and making fancy baskets of palm leaf, moulding rice cakes and stacking them in ornate designs. They were preparing for a temple feast.

A young woman separated herself from the group and came down toward us. As she stepped upon the little foot-bridge that crossed the moat, I stopped breathing. Have I said Renang was beautiful? Then surely this creature was a goddess.

Her skin was amber, unscorched by the sun, and her breasts were golden domes. She walked proudly, with unassuming haughtiness. Her hair fell down in a loose loop beside her face, light glittered on golden ear tubes, and her lips, falling apart in a calm deliberate smile, revealed the glint of pearly, even teeth. She was not a small woman, but her right hand had slipped through a tiny circlet of gold and tortoise shell, bright with pink rubies and Borneo brilliants, which seemed scarce bigger than a finger ring.

"It is the Anak Agung Sayu Putu Rapug," said Ida Bagus Gidéh. "She is the Old Lady's granddaughter."

Sayu smiled at us and spoke to Ida Bagus Gidéh. We could not understand them, for they spoke in Bali. Then she turned, and led us out into the palace courtyard and on through little silent houses into a tiny court, well-shaded with pomelo trees.

There on a chair sat the Old Lady, and maidens were combing her soft white hair. She smiled wanly, and motioned to a bench. Her breast was withered, but her back was straight, as a queen's should be. Her face was thin, patrician, and yellow like old silk. Hair floated around it like a cloud, as the maidens combed.

Roosevelt was talking to her, struggling against difficulties, for she had little Malay. But I was watching Sayu. She took a long pole, knocked down a pomelo. She broke its inch-thick yellow rind. She broke its inner pulp in segments like a grapefruit, and gave it to us. As a matter of fact it was a pretty poor pomelo, pithy and tasteless, but from her hand it was ambrosia.

Roosevelt was having a hard time with the Old Lady. She smiled and nodded amiably, but didn't seem to get his idea. Ida Bagus Gidéh began interpreting. The Old Lady was rather a disappointment to me, after Roosevelt's descriptions; she didn't seem very bright. But I didn't care. I was watching Ida Bagus Gidéh and Sayu.

Ida Bagus Gidéh, the handsome young man with two

homely wives. What a disappointment that romantic youth had been. But now the old thoughts sprouted up again. Ida Bagus Gidéh and Sayu. What a pair they were! Renang was a fragile bud of beauty, but here was full-blown loveliness. Was I an incorrigible matchmaker in my dreams? Well, never mind, Gidéh was a settled, married man.

The man writing facts is at a terrible disadvantage. The obvious, the expected, the almost necessary thing at this moment was for me to notice a drooping of Sayu's eyes, the flicker of a secret between Gideh and her. Fiction writers have an easy time. They take all the obvious, customary emotions and pantomime of white men, put them in oriental bodies and costumes, and have a story readers can understand. Now, whether there were any thoughts or emotions passing between Sayu and Gidéh, I don't know. But surely there was no sign of them. Apparently they ignored each other.

Roosevelt had given up. Either the old lady was just plain stupid, or she was dead set against photography. With another wan, friendly smile she rose, a frail, dusty, wisp of humanity; and followed by her maidens, she went within her house. Sayu, with a nod, went after them.

"Well, no picture this afternoon," said Roosevelt as we drove away, and then suddenly the thought struck me:

"Why, oh, why, didn't you make a picture of Sayu? She's worth a dozen of the old lady."

"Never thought of it. I was too busy seeing her beauty to think of photographs. But I will make her picture some day. It won't be good, though. You could never catch that regal spirit on a film."

Ida Bagus Gidéh broke in on the conversation.

"That old woman," he said, "eats two *ringits* worth of *chandu* (opium) every day."

XIV

RENANG'S little sister Silé, who was three years old and would start training next year to be a *legong* dancer, sat on the ground in Kumis's yard playing in the sandy dust.

With her forefinger she drew a large cross. Then, starting at the intersection, she filled in the sectors with a balanced, formal design. Gradually it grew to two feet in diameter, a simple but beautiful thing. A small naked boy came beside her. He picked up a fistful of dust and threw it over her drawing. I expected a quarrel, but there was none. She looked at him and smiled. Then together they smoothed out the sand and began making other patterns.

Silé was born with an impulse to create designs, but little impulse to preserve them. She was acquiring taste and a feeling for design unwittingly, as the child of a cultured home learns to use good diction.

Later that same afternoon Roosevelt and I were motoring along the road near by, and hanging from a doorway shrine was a palm leaf panel. It was about fifteen inches wide and six feet long, made of leaves which had been split to show a white surface and pinned together with splinters from the stiff skeleton of the palm leaf. Two of these were elaborate expositions of pure design. Also there was a charming *stylissé* figure of a dancing girl, with fan-shaped head-dress and long graceful arms, standing between two conventionalized trees.

I had seen many simpler palm leaf offerings, but had been looking for one of these; for Walter Spies had been telling me about them just a few days before.

"They make them only two or three times a year, at feast time, and then all the women make them, to hang upon their houses and temple walls," said Spies. "Last New Year's in Ubud I began to collect them, just one of each different kind. I soon had a great pile, and had scarcely begun to exhaust the varied supply."

He showed me exquisite inked designs he had copied from the withering palm leaf scrolls. Of each there were several variations. Each group had a generic name, and each variation had a name also. The making of them was the art of women everywhere, carried in the memory and passed on as a tradition.

When Roosevelt and I returned to our village resting place at sunset, after finding the offering by the roadside, we asked Réné, our cook, if we could have one made. We hardly got any dinner that night, for immediately a knife was flying in her fingers, and she was dropping fancy bits of palm leaf in a jar of water. That night Renang sat down to help her. The next morning, an hour or two after sunrise, I asked how the work was progressing. Réné brought forth the completed scroll. It was not quite the same as the other, for it was made in a different *banjar,* but it was quite as beautiful. Roosevelt photographed it, and a few hours later Réné threw it out into the door-yard, for the pigs to eat. Its intricate delicate refinements of pattern were already beginning to wither.

Ktot, who had done the carving in our *balé*, was work-ing in these days on the Padanda's gateway. The pillars of the gate had been built as two parallel piles of masonry from quarried blocks of what may have been sandstone, but more likely was dry volcanic clay. These were about ten feet high. The sides of the pillars facing each other had been squared off evenly, but the outside had been trimmed into a sweeping curve which flared out at the base. Over all this Ktot had erected a high temporary roof of grass, and now with knives and instruments like chisels was carving the front side of the gate. At the top of one pillar he had sketched the beginnings of an inter-twining flower pattern, and at the level of the eyes was coming into relief the figure of a weird man-like figure, a yard high. I knew what it would be: knees and arms akimbo, body in a warrior's tunic, sword in hand, glaring eyes and beak with fangs, in the background flaring wings —Garuda Bird. Ktot was working rather rapidly, goug-ing out the soft stone with his cutting tools. He wasn't following any particular design or model, was working things out as he went along.

When the gate was finished, the roof would be re-moved. Torrential rains would pour upon it. In a year it would be green and mossy crumbling. A few years more, it would be eroded as if a hundred years old. Then an earthquake would shake it down.

Late one afternoon Roosevelt and I happened into Bengkel to spend the night. We had just come down from Buleleng and were going some place else next day; where I forget. I went directly to our *balé*, but noticed that Roosevelt was going on with Kumis and Madé into the temple at the rear of the compound. In a minute Babad came running, and motioned me to come there too.

Now this temple was a rather disreputable looking place, unkempt and tattered. It is a pious act to build a temple, but rebuilding stores up no treasure in heaven. This temple had never had much decoration, and its few images were broken and eroded beyond redemption.

But there now stood Roosevelt and Kumis, and before them in a little *balé* sat an image, the like of which I had never seen before.

I had been sleeping and living with graven images for quite a while, and was getting familiar with them. Vishnu riding the Garuda; the *singa,* winged Hindu lion; the Rangda, wild-eyed and terrible; dainty ladies; strange beasts. The quaint assortment of little amusing figures, for which we could find no better term than "goops," some that might be perfectly at home on the Gothic façade of Notre Dame, some primitive as the simplest figures in American comic papers. I had acquired a few myself. I had a little primitive wooden mother, suckling a child, which everyone had been trying to get away from me.

But this before me was something different. A slim kneeling meditative figure with hands posed in Buddha-like meditation, its curving lines graceful as an Attic vase. It had a slender nose, and delicate modelled arching eyebrows. It was of the most severe simplicity. It was cracked across the base, but otherwise in perfect condition. It was not of wood, stone, plaster, or any other material I had seen.

"Terra cotta, I think," said Roosevelt, his eyes gleaming in exultation, his voice trembling. "Very old."

"A man brought it and left it here for me. He wanted a dollar for it. I gave it to Kumis. I was too excited to bargain."

We never found out what this figure was. Perhaps a Buddha, perhaps not. The natives could not agree. It was the creature of some forgotten past.

I have sat for hours contemplating its delicate placidity.

I was catching Roosevelt's enthusiasm for temples. Temples were not Bengkel's strong point. They had music and dancing in this village. Other places went in more for art.

It was against my will, on my first day in Bali, that I visited temples.

"Listen," I said. "I spent a day among the embalmed treasures of Nikko, and the only thing that gave me a thrill was the grove of cryptomeria trees. I saw temples all over Japan. I spent a day last week in Java at the ruins of Borobudur. They were magnificent and beautiful, but what I want to see now is the life of the people."

"Well, you'll see the life of the people," said Roosevelt. So we went to temples.

We went first to a temple in the town of Jagaraga. Its

disintegrating structure was apparently of great antiquity. It was a stone-walled rectangle, open to the sky. Steps mounted to a narrow door in a massive gateway, and over the tiny portal glared the huge face of Shiva's wife in her fierce incarnation as Durga—great staring eyes, white jutting fangs, a yard of tongue rolling from a crimson face. Right hand upraised, the traffic policeman of the soul. Not once but a dozen times in smaller figures, the ravenous visage glowered down from that tall gate. I entered.

The rectangle was divided into two courts by another wall midway, pierced by a roofless gate of tall columns, which was flanked also by narrow arches. All these were carved to feathery lightness, with incredibly varied flower designs, delicate, jutting, leafy cornices. The walls about us also were cut with flowers, with every panel different.

In the court beyond were shrines. A high flight of steps led to the central shrine, a blue box with gold trimmings and red tile roof. Beside it were two tall stone thrones, seats for the gods. The central shrine and the two gates were not in a straight line. For evil spirits travel in an un-curved path, and by this architectural trick they are ex-cluded.

About these shrines and inside the middle gate were images of untold variety. There were dignified knights and ladies of Hindu legend, there were more fierce demons. Amiable Membrayhut of Balinese legend, with a goodly delegation of her 108 ill-behaved children. The winged, man-like Garuda Bird, steed of Vishnu. A genial version of nourishing motherhood, supporting with her hands her vast long breasts. There were owls, bats, roosters, crabs upon the walls, from which also smiled down a friendly tiger. The throne shrines were mounted on the backs of great tortoises, about which coiled the *nagas*, the great serpent gods. And everywhere were fanciful designs of flowers and foliage.

On that first day I did not see this much of that marvellous effusion of creative skill and imagination. I was bewildered. I stood there cramped with visual indigestion.

There was an atmosphere of decay about the place. The images had all been painted, red, yellow, blue, black, and white, and apparently in recent years. But moss and weather had dulled their flamboyant hues. As I climbed a shrine, bricks crumbled away beneath my feet. There were shaggy, unkempt grass-roofed sheds in the courtyard. It made me feel a little desolate to look thus into the neglected past. Roosevelt was with me. As we emerged I spoke of this sadness of ancient decay.

"Look," said Roosevelt, and pointed to a bas-relief beside the outer gate.

There was an automobile, a ridiculous, burlesque automobile, with flag flying, horn blowing, headlight shining. In it were two long-nosed men with wild, frightened eyes. They looked down upon an evil bandit with a horse pistol.

"Look here on the other side of the gate," said Roosevelt.

There was a panel of outdoor sports, all intermingled. Two men in peaked hats paddled a canoe; a boy flew a kite; a fisherman hooked a whale. A comically Semitic

Arab in a fez rode a bicycle. Two airplanes battled in the sky. One plane was crashing, and two men on the ground pointed with amazement at the pilot, who had fallen overboard. The figures were carved with cunning crudity. They were the work of a man who had gone to the movies and seen them, not with a literal eye, but with the comic vision of a Rube Goldberg.

"There are a hundred men in this village," said Roosevelt. "Eighteen of them built this temple. They are just ordinary village men. They did the work without pay, as their part in community life. There is nothing of the neglected past about them. They are building a new temple."

The cave man who carves his newspaper on a cliff is a stock figure of the cartoonists, a standard bit of foolery; but here in Bali there is a veritable journalism in stone. I like the figure. The sculptor's materials are not much more durable than newsprint; his work is not given much better treatment than yesterday's newspaper; and much of it is comment on recent events. These bas-reliefs of Jagaraga are not unique. There are several stone bicycle riders in North Bali, and in a temple at Banjuning may be seen the grotesque man of the Michelin tire advertisements, running down a pedestrian. Next to him two soldiers with long European noses, who might be Sergt. Quirt and Capt. Flagg, hold a tug-of-war over a flat-nosed native maiden; while near by a brown-skinned girl accepts the embrace of a white man in an official uniform. In another new temple, a few miles away, the great panel behind the main shrine was entrusted to a young man who filled it with a hilariously realistic bas-relief of carousing European soldiers, doing great damage to an array of bottles. This has been cut away, and replaced with an innocuous design of flowers. A Government official thought it was not a fit subject for sacred decoration.

Not beautiful, but vital, is this comic supplement. While other brown peoples have cringed and aped the ruler, the men of Bali observe the ways of the white man and go their own sweet way, pausing the while to make amused comment on the idiosyncrasies of the foreigner. Theirs is a rich, prolific, vital art, which may be influenced, but will not be killed.

Bali's art is a religious expression, for the most part, but there is nothing sanctimonious about it. Religion and art are here too much a part of daily life for that. Anything decorative may find its place upon a temple wall. The Balinese are humourists. An amazing circus of familiar beasts and amiable clowns grimace from their temples. And especially is the comic spirit shown in those representations which in our smug English we are tempted to call pornographic or bawdy. These words are wrong. Done by a white man, such figures and groups would be smut. But in Bali, where insinuating sex-obsession is non-existent, a high-spirited *naïveté* purges them of all offence. Typically Hindu in essence, they have lost nearly all dogmatic significance, and are merely decorations.

It is, none the less, startling to encounter on a temple wall a chain-like frieze of grinning monkeys indulging in mutual, or serial, pederasty. It is gay to find, in a jealously sheltering shrine, a spirited interpretation of the lingam: a stone cylinder a yard high and a foot in diameter, with its round top carved into a brightly grinning, pop-eyed face, a figure with legs kneeling in order to trail upon the ground a weighty and affectionately sculptured member which has quite outgrown them. It impresses one with the possible irony of fate to see—as on the painted walls of Klung Kung's palace—gentlemen and ladies who have been most magnificently endowed for no other apparent purpose than to provide subjects for torture by the fiends of hell. And it is tantalizing to leave Bali, as I did, ignorant of the real purpose of those fastidiously painted parchments on which (in dozens of little squares) are represented all conceivable, and some inconceivable, permutations of solitude *à deux*. Were they (I still wonder) made to demonstrate sins to be avoided? Or were they (as one authoritative gentleman solemnly informed me) made as wedding presents, as the Balinese version of what every young bride should know?

At any rate, one thing is certain. These people not only dare to face the facts of life; they exaggerate them.

There is little that is pretty in this art. I imagine it would make no great appeal to Maxfield Parrish's public. I doubt if it would please much those travellers who gasp delightedly at the expensiveness of the Taj Mahal. I don't think it would give much pleasure to those Cairo tourists who pass by with a casual glance the solemn, austere mosque with which the Sultan Hammid justified his rape of the sheathing stones of the Great Pyramid, and who rhapsodize over the abortion which Mohammed

Ali made of alabaster. Bali's materials are shoddy. This art differs from ours also in spiritual and emotional qualities. You will find nothing here to compare to the shadowy dreams of Rembrandt or the lusty mirth of Franz Hals. There is no portraiture. It is an art of spirit and dignity, but more especially of design. No matter what may be his subject, the Balinese never forgets his composition.

I view less the excellence of individual pieces in Balinese art than the immensity of the whole art as a social and religious phenomenon. We westerners build museums and there store the souls of dead men. I have spent days of ecstasy in the Louvre, lingering before the living remnants of the great. But at times there came a horror of it all; here they were, Rembrandt, Rubens, Velasquez, ranged in rows and under glass, stacked neatly like corpses in a morgue. In Bali there are no museums: art is life, being born, replenishing itself, and dying, as much a part of this people as is their blood.

There is in the language of Bali no word for art. There is no word for artist. A man is a stone carver, a wood carver, a painter, a goldsmith; that his work will be a striving for the beautiful is taken for granted. The abstract idea of art, art for art's sake, is so ingrained that it needs no expression; and commercialization had never occurred to these craftsmen until tourists began making a demand for standardized trash. No work of art ever bears the signature of its maker. The author of an exquisite decorative panel which you may admire on a temple wall is probably a man whom you could hire for a guilder or a guilder and a half a day and put to work cleaning your front yard.

Not far from Jagaraga is the new seaside temple of Penimbangan, in which the makers triumphed over that unfriendly modern material concrete. One might think that temple at Jagaraga, which shows some Chinese influence, has too much rococo elaboration of detail. (I thought so myself. It cluttered my senses.) But this little temple of Penimbangan, with dainty little princesses and papaya trees, and not a hint of Chinese, is of austere simplicity. It is modern art, a purely Balinese and radical departure from precedent. One sculptor triumphed especially with a girl, with water pot on her head, under a tree—all imposed on a cobble-stone wall. In enthusiasm one day, Roosevelt asked who the maker was. A hundred yards away he found the artist, labouring knee-deep in the mud of a rice field.

But the artist is respected. Skilled craftsmen of olden times have been enshrined as minor gods, and it is not unknown for a man to be made Pungawa, chief magistrate of his district, solely because of his artistry. In the old days the Rajas gave some of their artisans rice fields, to subsidize their labours. Today their descendants live from the produce of the same fields, and carry on their fathers' handicrafts. There are villages given over almost entirely to the practice of certain arts, painting, the working of gold, the carving of wood; and there are artists in virtually every village. And of course they do most of their work without pay, as their contribution to the community life.

There are among the million people of little Bali, I doubt not, more men entitled to the name of artist than in all our populous states. I was talking about them one day with P. A. J. Moojen, a Dutch architect and amateur painter of Java, who has written a book on the art of Bali and probably has done more to preserve its art life than any other white man. He told me of the days after the 1917 earthquake, when for the Government he assisted in the repairing of the Raja's palaces; of the traditions of art carried on for generations in families and villages, of the absorption of men in the making of beauty for the glory of the gods, of the pride of older men in the development of their pupils.

"Tell me," I said. "You know much more of the world's art than I. How would you rank this people as artists?"

"Why, there is no question about it," he said. "As a people, they are the greatest artists in the world."

Three other painters were in Roosevelt's house with us at the time, and I heard not one word of dissent.

That temple in the town of Jagaraga was one of a half dozen. Every town in Bali has its temple, and most of them have several. Most of them are decorated. The artists are the ordinary men of the town. There are few temples in Bali more than sixty years old, and very few lasted through the great earthquake of 1917. This very circumstance keeps the art of Bali a lively thing. There are no classics, save a few images enshrined as sacred antiquities, and Bali cannot rest on the laurels of a few men in a bygone age. The artistic efforts of the whole population are needed to compete with the ravages of time. There is no effort to repair old temples. But always is being replenished this art which presents itself at every turn of the road. It is an art which is not the possession of any Greenwich Village, or Fifty-eighth Street, or Bohemian quarter; it is the avocation of these men you see sitting along the side of the road, tending their fighting cocks. Some of it is bad, some is good, and much of it is exquisite.

XV

ONE night I stood on the beach at Sandakan, North Borneo, throwing pebbles in the sea. For four full days the rain had poured, and this night, though dry, was dark and overcast. The dropping pebbles startled the water into brilliant flashes of incandescence. The sea was vibrant with phosphorescent life. Had the night been windy, mariners would have been blinded by the water's glow.

A pinch of sand, tossed from my fingers, burst in the water like a rocket; and then the night was dark again. The flaming wonder of that evening seems incredible in memory.

As I look back through intervening months to those later days in Bali, it also seems incredible that ever I was so much alive. Thoughts, emotions—could life ever have been so phosphorescent? I lived in a strange world, on the brink of unsuspected wonders. A touch, and the world was flaming, illumined with a prophet's flair. I stood at the gate of wisdom. Then all was dark again.

After a week in South Bali, I would return to Buleleng and try to write, try to capture some faint glint of ecstasy.

Hours I sat motionless at my typewriter, impotent before the wonder of the world. Something was here, something profound, could I but capture it.

Roosevelt would go to bed, and I would sit there pondering. I would fling open the door, stride out into the glowing night. On such a night I came in, and rapidly I wrote:

"I came from the Isle of Manhattan to the Isle of Bali. I came from the land of health and wealth and tenements under the Elevated. From the land of the Individual's importance. From the land of liberty, equality, and opportunity, where a man screws on a particular bolt in a particular way several hundred times a day for a billionaire, and thus earns leisure to read the advertisements and learn that he wants a seven-tube radio and an electrical refrigerator, and that by dutifully buying them (on the instalment plan) he can speed up production and make opportunity for himself to screw on 300 bolts tomorrow where he only screwed 200 yesterday. I came from the land where millionaires endow music and art, where Irving Berlin and Bud Fisher make millions. From the land of Canon Chase and Bernarr Macfadden's board of ministers, where sex is sinful and rigidly expunged from everybody's mind, where Gilda Gray wins fame and fortune. The land of ambitious struggle, progress, speed, where in the shrieking of Park Row you can buy tomorrow morning's newspaper tonight and thus lose no moment in reading of the good, the true, the beautiful. All these and much more I had, in the rings of white supremacy's great circus; and yet, strangely, I was not satisfied.

"I came to a land beyond the blessing of the Gospel, where men whore after strange phallic gods, in shameless nakedness. Where skins are brown, hence obviously inferior. Where dance is disembodied line and rhythm, and where in music springs no lust or passion. Where the individual means nothing, is a mere cell in a social organism, who lives as he pleases and in his anonymity blends his soul with art. Where men know naught of passion, love, or hatred, and die as they lived, as tranquil drops in life's immortal stream of protoplasm. And is it plain that here I found surcease from cheapness, that here I found a peace which passeth all understanding?

"But now I seem to sense a faint response of incredulity. Why, then, you wonder, do I return? For I shall return, and while you read this I shall be religiously striving to get the better of my neighbour, and with laudable ambition coveting my boss's job. And though I render lip service to Bali's cerebral counterpoint, it is quite probable that one of these evenings I shall go past Town Hall where Harold Samuel is playing Bach to a small and select audience, and shall crowd into the Metropolitan Opera House, just in time to revel with sentimental abandon in Bori's second-act aria, *Connais-tu le Pays?*

"You see, it's this way. I am white, somewhat Nordic, with a tendency to blondness. That's my hard luck. The peace of Bali is for brown men. But I am white, and in me there is lusting turbulence; it is for me to assume the burden of the white man's tradition, his supreme civilization. It's my way and I can't help it. Oh, I might seek heaven here in Bali; but this I know, that if I did, I should go plumb to a white man's hell, without the inconvenience of dying.

"The peace of Bali is for brown men. And this I know is the white man's burden: that he shall dream dreams, and they shall mock him, that he shall seek what he cannot find, that in him there is lusting turbulence, and for him there is no Nirvana. He wants, I want—peace? Oh, I am not done with doubting, and cannot take what I cannot understand."

In the morning, when I had slept, I read this over. Did it mean anything? I wasn't sure.

DEMI-MONDE

XVI

WHEN Roosevelt and I came back to Buleleng from our sallies in South Bali, the *babu* would look after us. She would come out to the gate, bursting with laughter, and carry in our baggage. As a servant, the *babu* was just about as haphazard as the household in which she worked, but I got pretty fond of the *babu*.

Still, she had to be sent away. She kept things quiet and peaceful while she was there, for the most part; but, paradoxically, I never realized until she was gone what a little termagant the *babu* was.

She was an ugly little thing. Javanese, and flat-chested, with a tiny turned-up nose. Her *bajus* had a way of getting all twisted and slattern, and her hair would straggle down around her face, which was usually puffed out at the cheek with a wad of *siri* or tobacco. Oh, she was ugly all right.

Yet sometimes she was pretty. In the mornings she would wash clothes, on the concrete floor outside the plunge bath. She would wear only her *sarong,* hitched up over her breast and hanging just to her knees. Her hair would be wet and slicked back into a little knot. The sun would strike down through the mango trees, upon her smooth brown shoulders. She had dainty, pretty arms and ankles, too.

The *babu* would laugh a lot in the mornings. She had a high-pitched childish voice, and childish laughter. She would burst out without warning into a peal of treble glee, and just as you were wondering what sort of imbecility this was, you would find yourself laughing too. She would laugh in the evening when the tokay cried. The tokay was a lizard about a foot long, which lived out by the bathroom. "Geko-o-o," he would cry, in a stentorian, guttural bass. "Geko-o-o, geko-o-o-o, geko-o-o-o-o." Each time a little softer, deeper, slower, as his wind ran out. The *babu* would listen, breathless. At the seventh cry, her face would burst into a smile. And if the tokay gekoed nine times (as this very excellent tokay usually did) she would break into a tremendous flutter of happiness. Oh! that was good luck!

The *babu* took good care of us. If you ran out of cigarettes, all you had to do was to cry, *"Babu, rokko ada?"* and she would dig out somewhere a fine cache of Mr. Bear's Elephants. Or *"koré api ada?"* and she would come running with matches. Or *"ys ada,"* and she would go running down to the ice plant with the vacuum jar to get two pounds of cocktail cooler.

She could make good coffee. She always had it the first thing in the morning, and at four o'clock in the afternoon. And you who think that Java is a synonym for coffee are quite mistaken. Here is what they do to their coffee in the best hotels of these islands: They roast it black, then grind it to a powder; at night they put it in a filter with hot water, to produce in the morning a thick, cold, black "elixir"; they mix this fifty-fifty with a hot dilution of canned milk; the result is a tepid, nerve-shattering cup of nausea. But our *babu* had learned to roast coffee, and boil it right, in an old aluminum pot with the handle half twisted off.

The *babu* was a good cook. She would make a fine *rijst-tafel*. Perhaps you've heard about the rice table the Dutchmen eat. Your boy brings you a big bowl of rice, from which you help yourself generously. Then, if you are in a big hotel, and turn around to look, you will see behind you a line of perhaps a dozen boys, each with a dish or several dishes. They are presented to you one at a time, several kinds of chicken, beef, fish, various pickled and stewed vegetables, coconut and other relishes. When the first boys have served you, they go get other dishes and take their places at the end of the line. After a while you have a great stack of cold and somewhat clammy food in front of you; and if you are a Dutchman you attack it with both hands until you've eaten just about all. Then you go and sleep for two hours. The *babu's rijst-tafel* wasn't like that. It came hot and savoury from the kitchen, and there was just about enough for a gentleman's meal.

The *babu* gave us duck's eggs for breakfast, and often there were ducks for dinner. She had a flock of ducks that she kept in the back courtyard. Balinese ducks do not waddle; they stick out their chests and strut. There was nothing to bar the way, so two or three times a day this band of ducks would come marching out into the centre court, like a troop of Mexican generals. "*A do!*" the *babu* would cry, clapping her hands. "*A do!*" And those intelligent ducks would turn right around and march back again. The *babu* had bought her ducks for twenty cents in the market, and she sold them to us for a guilder; so we had lots of ducks to eat.

All in all, it may seem that she was a very satisfactory *babu*. And she was, except for her tantrums.

One afternoon Charles Sayres, a Dutch artist who made headquarters with us, was painting a rather bad picture of a very pretty girl. Roosevelt and I were making photographs. Sirone and Nioman were posing for us. After a while we sent the *babu* for some ice, and we all had beer. A very gay time we had. We offered the *babu* some beer, but she wouldn't drink any. She had been staying out in her kitchen all that afternoon.

Now she came to gather up the glasses, nobody paying any attention to her. I happened to notice her as she walked across the courtyard. Suddenly she stopped, wheeled about, flung a tumbler across the cobble-stones. The glass flew to bits that shimmered in the sunlight. She threw a plate, that burst to tiny fragments. She stood there, her face contorted, her body trembling in every limb. Then she went out to her kitchen and stayed.

I shall not attempt to explain why the *babu* acted this way. I am no expert in Javanese psychology. Was her pious Mohammedanism shocked by a lot of brazen hussies sitting around with their shirts off? The *babu*, as I have said, was an ugly wench. Nobody ever made her picture. Was her vanity outraged? All this was speculation. But one thing was certain: painters and photographers must have models. And also it wouldn't do to have a Malay running *amok* among the crockery. There was little enough crockery as it was. When that was gone, what would prevent her from starting on the precious Sungs?

"*Babu,*" said Roosevelt when the girls were gone, "if you ever act like that again I shall send you away."

The *babu* stood sullenly, in red-eyed anger. Then she went back to her kitchen and started packing. She walked back and forth through the house, gathering up her possessions.

"*Babu,*" said Roosevelt, "if you don't get our dinner, I won't give you any money."

So the *babu* got our dinner, then went on with her packing. It was amazing, the number of things she had stuck away in various corners of that house. A pair of sandals here, a *sarong* there, a piece of matting somewhere else. She made a separate trip for each article. Then she went out in the back yard and began gathering up her sleepy ducks. She tied them together in bunches by the feet. Out by the bathroom the tokay croaked six times and stopped. Our bedtime came, and she still was packing. She stood there, with a bundle in her hand.

"I'll come back in the morning and get the rest," she said.

So we went to bed, wondering what we would have for breakfast.

I was awakened by gay laughter. I opened my heavy eyes and saw the *babu*. There she stood, bright and childish as ever.

"Good morning, *tuan*," she cried. "Coffee."

The storm had passed. Everything the *babu* owned had been put back where it was yesterday. Chummily beside the plunge bath lay the *babu's* bunch of *siri* leaves, keeping cool and wet. Out into the courtyard strutted a pompous and possessive troop of ducks.

But it was just a lull. A few nights later we were having a party. A couple of Dutchmen were there. The chief of police dropped in. The chief was a very good cop, I think, for he never appeared except at the precise moment when a bottle was being opened. Sayres had quite a number of models on the string, and during the evening he slipped out and came back with a few of them. We were having quite a genial time.

Then suddenly there was a commotion out by the kitchen. I just saw a flash of it. The *babu* was on the war path. She had little Nioman by the hair. Then Nioman got away, and slipped out by the side gate.

That finished the *babu*. She packed in a hurry this time, and was gone. We were wondering how we would be fed, and then it was a miracle happened. The whole neighbourhood moved in on us. Everywhere you looked there was a native, working like mad. The yard was swept, the floors dusted, clothes washed, a feast prepared. The *babu* had had a good job. It didn't pay much, but there were always visitors in this house; and that meant presents, prosperity. The *babu* had had a good job. She had guarded it jealously, fearsomely. She had kept the neighbourhood in terror, little Tartar. But now the bars were down.

Everything would have been hopeless confusion, if it

hadn't been for Sironé. She took charge. A remarkable person, Sironé. High-spirited, noisy, ribald, entirely out of tune with the peaceful imperturbability of Bali. Yet she was Balinese. She had an ugly, pock-marked face. But her teeth were right out of a dental advertisement, and her body was as beautiful as any in this island. I guessed she was about twenty-five. Beauty blooms and fades early in Malaya. But Minas told me I was wrong in this case. Sironé, he said, had looked exactly the same for twenty years.

Sironé had been married lately to a Chinaman. The Chinaman had had a motor car, and had taken Sironé down to the feast at Bésakih. Then the car broke down, and Sironé had to walk home. She left him after that. She may have looked the same for twenty years, but during twenty years in a white man's town she had learned a thing or two.

She was a good cook, and a good manager. The last I heard, Roosevelt had made a moving picture actress out of her.

XVII

A MAN from Tahiti came to Bali. The man from Tahiti was stroking a brown girl's silken hair.

"The girls of the South Seas are gentle, friendly things," he said. "Their little tendernesses are very pleasant."

This brown girl was no better than she should be. If she had been, he probably wouldn't even have been playing with her hair. I was surprised that she let him do it, anyway; for she was a matter-of-fact person. She sat there for nearly half an hour, while he stroked her frangipani-scented locks and looked upon her mooningly; but she didn't get very much interested in the proceedings. Though she might have been amiable, she was not ardent.

Finally he rose, and picked up his hat. He spoke judicially.

"Bali is a wonderful place, and the natives are charming," he said. "But I think I like Tahiti better."

And he walked out into the passionate night, alone, toward the rest-house.

The brown-skinned girl was laughing at his back.

When I tell Americans about Bali they almost invariably, in one form or other, ask me one question: "Are the people moral?"

I have been puzzled somewhat in trying to answer that, and the best reply I have found is another question: "Are Americans moral?"

My friends in Bengkel, if I were telling them about my island, would never think of asking such a question. I don't think the matter would interest them, particularly.

One day in Buleleng I was packing a box, and I sent a boy down to the Chinese shop to get me some newspapers. Now in the Indies the American newspaper has reached its ultimate usefulness. Bales of them, mostly ancient copies of the Los Angeles *Times,* are shipped about for wrapping paper and used in the Chinese shops. This day, on top the pile of stuffing paper the boy brought me, rested a complete and unread copy of the New York Sunday *Times* for Feb. 5, 1928. What wrought this miracle of news from home I never knew, but I fell upon it like manna in the wilderness. And there in it was Mr. J. Brooks Atkinson's article on Eugene O'Neill's new play, *Strange Interlude.*

In this strange interlude of mine, nothing seemed farther away and stranger than the mind of Nina Leeds and of the people who wallowed in her neurosis. I think I have made clear how alien sex preoccupation was, in this land where Bernarr Macfadden would go bankrupt. But since my American friends have been reading South Sea bunk, I may as well answer them about Bali, which I emphasize is about 5,500 miles west of Tahiti.

Not long ago the parents of a girl in North Bali, under an old native law, had her sent to jail along with two young men, because she had been too complaisant. In the old days, a man taken in adultery was put to death and the offending woman was appropriated for the Raja's household—which possibly indicates that a certain value was put on these rare ardent ladies. The native religious law is very strict about extra-marital relations; and in North Bali (except for those young ladies who always appear to supply a demand, without amateur standing) the people are inclined to follow it. In South Bali the standards vary. Young people generally do not marry until they are in their late teens, and there are many bachelors at twenty-five; but it is doubtful if many young persons marry without previous experience. In some localities sex life begins at puberty, with more or less common knowledge. But there seems to be little illegitimacy. Perhaps there is knowledge of birth control; some of the most primitive tribes among these islands practice contraception. In Bengkel the standard seemed to be one of outward order and decency. I never saw a sign of spooning, but I know there were intimate couples among the *banjar's* youth.

There is no more appropriate place than this, probably, to mention that in Bali, as in all oriental countries, there is a certain amount of male homosexuality. Of this I saw no overt indication, but I am told that it is widespread and that no especial moral stigma attaches to it. I often wonder what Katherine Mayo, and other such favourites of the Ladies' Aid, might feel impelled to write about Bali; but I know that to emphasize peculiarities of sex life would be to tell falsehood by disproportion. I also have wondered what would happen if the Orient were not oriental, what might be the population of India and China and how many more persons might starve to death, if Nature were always natural. As a matter of fact, in a climate which makes the sex instinct especially strong, and in countries which practice polygamy, it appears almost a mathematical axiom that male homosexuality will be wide-spread. In Bali, for instance, women outnumbered men in 1920 by only about 14,000; obviously there were not enough wives to go around twice. That homosexuality is condoned is not to be taken as indicating that there is in Bali no moral sense regarding perversion: cohabitation with a beast is an offense punished by exile.

I know of not one Balinese half-caste, although the whites have been in North Bali a good many years. This is especially remarkable, since in near-by Java there has been such infiltration that mixed blood is scarcely a social handicap. Among the 250 listed as "whites" in Bali's census, well over half have Javanese blood; most of these are Government employees, officials, or more generally their wives. Balinese blood, too, will undoubtedly become mixed, before long. White men in the tropics are inclined to have their mistresses, in an unromantic, matter-of-fact way. But this in Bali is no casual affair.

"Because the people of this town know me well and like me, I could have a girl or girls without the slightest trouble," said one man, who lives close to the natives in a town where amorous relations happen to be the most free, easy, and oriental in Bali, but who does not happen to care for femininity. "But with a stranger it is different. For the travellers who sometimes stay with me a day or so, for instance, it would be ridiculous to try to provide such hospitality. It would be impossible."

I've never been to the South Seas. Maybe they are a garden of romantic dalliance, as they say. In Bali there is no more need for being celibate than in any other country; but so far as Bali is concerned, the man who wants to be somebody's darling, who yearns for languorous language lessons on the beach, had better stick to his South Seas.

XVIII

"I WANT a bed," said Plessen.

It was a Friday morning, and Plessen had just come ashore from the weekly boat. He oversaw the landing of his car from a *prahu,* an old automobile with a snaggle-toothed differential, and then came up to Roosevelt's house. He brought Per Lund, a young Swede who had come out from Europe on a freight boat. Plessen had met him on the boat the night before. "He can paint," said Plessen. "I've heard of him." So Per Lund came to stay at Roosevelt's house too.

I came yawning out of my room, hitching up my sleeping *sarong.* I hadn't had my coffee yet, but curiosity banished drowsiness. I looked this man Plessen over carefully. A legendary figure. I knew about him, but I hadn't quite believed in him.

Victor, Baron von Plessen, late of the Kaiser's Guard, had launched his career as a pup lieutenant just in time to parley with revolutionists for a doomed regime. A tall blond youth, not yet turned thirty, he stalked up and down the Indies, hobnobbing with Governors, giving the glassy stare to officious Dogberry *controleurs* who yapped at his contemptuous heels. Here was a man who had spent eight months in the jungle, away from any white face, and glad of it. Lord of vast acres in his own land: but why be a big butter and cheese man in Schleswig-Holstein when there were wild beauties to be painted in the tropic seas, strange customs of brown men to be learned, flamboyant birds to be taken for the Berlin Museum? Son of one of Prussia's greatest families, he mingled with the humblest coolies as a friend. *Noblesse oblige?* The ruling Dutchmen were puzzled, and called him socialist. The natives called him *Tuan Raja.*

Here, in fact, was the man who had sailed up to the remote island of a virtually independent Sultan, and intimated by messenger that he would receive the ruler aboard his boat. Who, when the Sultan had respectfully come aboard to call, had calmly indicated that he would be glad, during his stay, to occupy the ruler's house if the ruler would move out. And who had got the house.

"I want a bed," said Plessen. "A real bed with springs."

Now, to appreciate the enormity of this demand, one needs a better idea of Roosevelt's ménage. He had a tolerably large European-style one-story house, with three bed chambers, a dark room, and a great three-sided room opening upon a courtyard, beyond which were the kitchen, bathroom, and servants' quarters. The place was somewhat furnished. If you wanted to sit in another room, you carried a chair in there. The narrow beds were serviceable. After the first few nights, I learned quite efficiently to dispose odd anatomical corners between the slats. I would never have thought of asking for a real bed with springs. I wouldn t have got one if I had asked.

We started talking as we ate our breakfast of ducks' eggs, and talked all morning. Plessen, who I believe had never been west of the English channel, talked the American idiom with an English accent. He had just come down from the Jieng Plateau, Java's great table-land of Hindu ruin. Temples of Brahma, images of Shiva and all the pantheon, master works of ancient craftsmen, tumbling in lonely, frigid silence. He had paintings to show us.

Plessen had just come from the Sultan's Court at Sura-karta. They had had the *wyan wong,* the dances of the shadow-plays performed by humans—survival of Hindu-ism in a nominally Mohammedan Court. Three years the Princes of the blood had practised, for this one day's dance.

"I never had believed that human art could reach such quintessence of perfection," he said, his voice trailing off in the spell of memory. "Every finger-tip, every gesture— And the ringing of the *gamelan*—"

An amusing thought struck me as I listened: this was the barbarian Hun whose frightfulness the American Legion went out to stop.

"I stayed there from seven in the morning until eleven at night," Plessen went on. "My food was brought to me. A great many of the official Dutch were there. They had bridge tables set up in an ante-room so they could be amused."

"Swine!" muttered Roosevelt.

"There were two Dutch women next to me, who were

always laughing and talking. Finally I said: 'I'm going over and sit with my equals!' The rest of the time I was with the royal Javanese. Fine gentlemen they are."

As we talked, a miracle was happening. The place was overrun with natives. And there before us was arising an architectural wonder of white steel. There was a bed! A great wide bed with springs, a great, thick mattress, snowy sheets—complete with "Dutch wife!" *

I am sure no respectable Dutch *frau* had given up her couch for the convenience of any of the nondescript demi-monde that drifted in and out of what we sometimes called the Hotel Roosevelt. Had some brown-skinned neighbourhood Delilah turned temporarily from the path of sin? I do not know. At any rate, there was a bed. A poem of a bed. An ode to Morpheus. A glorious bed, from whose tall pillars floated a filmy, lacy cloud of net.

Tuan Raja was back.

* See note on p. 288.

XIX

THESE were carefree days in Roosevelt's house, but there was thinking too. Here was refuge for contemplation, out of the dazzling world where every moment was stunned with the impact of surprise. Here, in this lazy life, with nothing much to do but sit and think, I was thinking about Tintiya, god above all, whose only function was to exist. I was thinking of religion, which was the mainspring of all that I had seen. Especially I was thinking of what I had seen at Besakih.

Christians erected their great cathedral to Our Lady of Chartres on the site of the Gauls' ancient shrine to their own God the Mother, and the Hindus of Bali have built their national temple on the slope of the Great Mountain. Its towers are the pagodas of Shiva, but men pray there to the God of the Mountain, who was a god before the Hindus came. Every town has a temple at Besakih, and in their midst is the great temple, with massive flights of steps and tall pagodas. On a day in April Bali makes pilgrimage there.

A narrow trail leads across a wide sloping plain to the

mountain. I stood there and watched them coming, a long long never-ending line, men, women, and children, swinging up the slope. The women's heads were piled high with offerings, and they swung along with erect, easy grace. Their arms were swinging also. The rain poured down, and glistened on their breasts. Up near the temple an orchestra tinged the air with brazen counterpoint, and masked dancers wove their arms in grotesque pantomime. Under thatched roofs long-nailed priests made fantastic signs with their hands, rang their tinkling little bells, and sprinkled holy water. Shrines were heaped with rice cakes and fragile, lacy offerings made of palm leaves. Men and women knelt before shrines, put holy water on their lips, and flower petals on their foreheads, and then they went away. Under foot was trampled a myriad of palm leaf patterns, which had served their purpose.

And up across the plain they still were coming, with their easy tireless stride. Thirty thousand came that day. Ten, twenty, fifty kilometres they had walked, with offerings on their heads. And as the light began to dim, they still were coming, through the rain.

There come to a man strange, inexplicable moments of exaltation. One such came the day I stepped into the Sainte Chapelle at Paris. I stood suddenly in a place of delicate arches, all ascending, reaching upward. And all the world was spangled blue and scarlet, from a wealth of ancient glass. "God!" I exclaimed, in my somewhat common habit. Then all at once, sheepishly, I knew that I had spoken the one word that expressed it all.

Well, there I stood on the plain of Besakih, drenched

and sticky; and such a moment came. I was seeing something that had perished from the earth, with the Crusaders, with the men who lifted the great Gothic spires; for here in Bali a whole community, with perfect unanimity and sincere simplicity, was reaching out, however blindly, toward its aspirations. It was all very wonderful and strange, and it puzzled me. What was it these people had which we had lost?

I came from a country that sends missionaries to the heathen, where sects divided fight among themselves, and preachers prance as mountebanks to rally dwindling flocks. We had conquered the air, the ether, and the molecule. We had learned science, learned to put the microscope to facts and say: this thing is true, but of that I have no proof. Thus we had won wealth and power and magnificence. But we had lost something, something beautiful. I had known it when my throat thickened at the sound of bells on Christmas Eve. These people of Bali had it. Why? It all seemed very complicated, somehow; yet it was all so simple.

I was talking to Plessen one day about this matter of religion. We were sitting there talking, for there seemed nothing else that we could do. Per Lund was in the front room, painting a nude of the cook. Sayers and Moojen were in the back courtyard, burning their canvas with the line and colour of a dancer's robe. A Chinese carpenter was nailing boxes on the front porch, and out by the bath the *babu* was washing clothes. Plessen had spent an argumentative morning overcoming Roosevelt's aversion to the strenuous life, and Roosevelt had retired to his dark room to develop Plessen's photographs. Only now and then would he burst forth to vent his opinion of warm water in developer. Plessen's efforts had exhausted him, and in this atmosphere I had decided not to try to write. So we were just talking. And as the life of Bali is religion, we were talking about religion.

"Plessen," I said, "you know these people. Painting, exploring, you have lived among them. What is the hold of their religion on them? Everything they do is religion. Shrines in every rice field, shrines in every home. The other day when we were going to make pictures of the dancing girls at Bengkel, they were late, and I went to call them. There they all were, in their costumes, making offerings to the gods, and I too was sprinkled with holy water. How can religion so permeate a people's life?

"They have something which has vanished from our Christian life. I'm not speaking of myself. I no longer bend the knee, I guess I'm hopeless, but I have pious friends. Did you ever notice the sneering smiles and frowns of good Christians when it is remarked that some young man is going to enter the ministry? But here the priest is honoured. Men serve the gods in everything. What is the reason for it all?"

As I say, it all seemed very complicated to me. But Plessen had spent eight months in the jungle. He had just come down from the silent places of the Jieng Plateau.

"They believe," said Plessen.

XX

"I *WILL* get a tiger! I *will* get him!" cried Sayres, rushing in from West Bali. His eyes were red and wild. He was a bundle of nervous tension. He was dashing around.

"Three days and three nights I sat in that tree, without a bit of sleep. I could hear the tiger in the bush, walking and grumbling, but he would not come near my bait. And last night I dropped off for just a wink of sleep, and he came and took that meat away!

"I *will* shoot a tiger! I *will* get him!"

And with that, Sayres subsided into his bed and was heard of no more until the next morning.

Except for his shoes and striking European features, you would have taken Charles Sayres for a native. He was bronze from the waist up. He never wore a shirt unless he had to. He never wore a hat. Sayres was the son of a Dutch plantation man in Java, and an uncle of his had been an official in Bali. But he was a painter and liked to stay in native villages, so the Dutch society of Bali didn't pay much attention to him. He made his headquarters with our Bohemian lot. He had come as an excitable, temperamental painter. But now he was an enthusiastic huntsman.

Sayres had become a hunter somewhat by accident. When he had come to Bali, some months before me, his father had given him a .50 express rifle, a veritable cannon, an elephant gun. Sayres had fired it a couple of times to feel the kick, then put it in a corner and forgotten it.

But one night, when Sayres was in the house alone, excited natives came by.

"Tuan Sayres! Boya di laut! Boya!"

A crocodile? A crocodile in the roads of Buleleng? Impossible! Unheard of! None had been seen there for twenty years. But Sayres was all action. He rammed five shells into his field piece. He snatched up his flashlight. He strode down toward the seashore. He focussed his searchlight to a penetrating beam, and flashed it on the water. Sure enough, there in the smooth sea, a hundred yards away, floated the long, ugly snout of a reptile, with a glassy red eye.

"Hold the light and stand behind me," Sayres told a boy. He took careful aim. A thunderous shot shattered the placid night. The crocodile vanished, with a terrific swirl.

They leaped into a *prahu,* and paddled out to sea. Ah, the water was red! Now they could see the croc, down in the quiet clear water. A giant, thirteen feet long, and completely dead. Sayres had shot him, precisely through the eye.

With rattan ropes they hoisted the beast aboard. Sayres sat by proudly as they started to paddle ashore.

Then, with a last, terrifying reflex, the crocodile switched his tail. Sayres was on his feet. His gun was at his shoulder. He volleyed and thundered. Then he began to bail.

They got ashore all right, took tally on the marksmanship. With five shells, Sayres had put three holes in the crocodile, and four in the boat.

Now that he had smelled blood, he was a confirmed huntsman. Art suffered.

While Sayres slept off his tiger hunt's fatigue, we dined on venison he had brought. It was not dry and wild-tasting, like our American venison. It tasted like veal. No, I don't mean that it was veal. In the morning Sayres was up early, gathering his supplies for another invasion of West Bali. He took Per Lund with him.

"I *must* get a tiger! I *will* get him!" cried Sayres as he started out. I do not know how long he sat in the tree this time, for I had sailed homeward before he returned. I know he sat there a long time, for he is Dutch, and pertinacious. I have heard later that he abandoned his tree sitting and tried hunting the tiger by the scent of his own nose. And I can report he did not get his tiger.

There are tigers in Bali, though. There were, anyway. Through the narrow straight which divides Bali from Lombok, the next island to the east, runs an imaginary line which is very important to science. From Lombok eastward is biological Australia. West of the line is biological Asia. In Lombok's jungles are wild cockatoos, but Bali has none. Lombok has no tigers, but they are native in Bali.

Vardon once killed a tiger in West Bali. West Bali,

Jembrana, is a narrow strip of land running off toward Java. It has no high mountains, and so is rather dry, mostly neglected jungle. Vardon came into Buleleng once in a while from West Bali. If you gave him a drink, he would tell you about a tiger. If you gave him two drinks, he would tell you about two tigers. For a bottle, Vardon would trade you a whole menagerie. Vardon would look the visitor over appraisingly: "Got any money? Come out to West Bali and shoot a tiger." Sayres was doing his tiger hunting out by Vardon's place. Vardon had killed a tiger and had a skin to prove it.

Vardon was a long, brown Armenian with a drooping moustache and a hearty voice. Once he had been a millionaire. Twenty years ago he had been the moving picture king of the Indies. The strange new white man's show went like wild-fire among the natives, and their money rolled in on the pioneer Vardon. He never knew how much money he had. He would go to his ticket seller, and demand money, and fill his pockets from the cash box, and call his cronies to a party. So he never knew how much money he made; or how much his ticket sellers made, either. But once he had at least a million guilders all at once. Now he lived in West Bali, with his nephew, Minas's brother, on Minas's farm.

There was something queer about this farm, which I never fathomed. No one but natives had a right to own land in Bali. But in barren West Bali Minas had a cattle farm. But then, Minas was an Armenian.

XXI

MINAS was likely to drop in at four o'clock almost any afternoon, for coffee. Jacob Methusaleb Minas. "Spell it with a *b*" said Minas. "It's out of the Bible."

Minas was likely to have on his pyjamas. This was quite proper. They are very particular about dress in the Dutch Indies. If you turn up in Java wearing short pants, in the British fashion, you will shock the solid white citizenry. They will think you very ill-bred, and probably will tell you so. Plessen and Roosevelt were quite *declassé* in Bali, largely because they showed bare knees. But if you put on your pyjamas in the afternoon, and sit out on the lawn or on the steamer deck, that is quite *au fait.*

"Have you got coffee?" Minas would say, and sit down to talk. He supplied the coffee, which he had aged four years. In a Malayan sing-song, Minas drawled a bizarre English idiom. He was born in Persia, educated in Calcutta. At least he showed his schoolmaster a brawny and pugnacious arm, and stayed in school until he got tired of it.

Minas had come to Bali twenty years ago, working for Vardon as a motion picture operator. He travelled on foot, up and down Bali, with his projector and films on the backs of coolies. Soon he decided to be his own boss, and paid Vardon an outlandish price for his ramshackle outfit. Then when films came C. O. D. he had no money. But he struck a deal with the steamship agent, and paid for the films after his first night's show. After a while he settled down in Buleleng. The Balinese were a thrifty lot, and after they had seen the movies once or twice they would not go any more. Paying admission to a show was a custom quite foreign to them. But in Buleleng there were enough Mohammedans and Chinese to make an audience. Minas did some trading, too, in coffee and cattle.

One time Minas came back from a trip to Java, and he started up to Munduk to get some coffee he had bought before he went away. He felt very much pleased as he drove up toward Munduk, for coffee had been skyrocketing and he had made a nice profit. The coffee woman at Munduk met him with a smile. "Tuan Minas, I have good news for you. That coffee of yours was starting to get mouldy. But I sold it for you, at just the price you paid for it, and here is your money back."

"So-o-o," said Minas, as he sat down and drank a cup of coffee with her. "I came to get my coffee."

"But I told you, Tuan Minas, that I sold it for you and saved you a lot of money. Here is your money."

So they talked, and drank several more cups of coffee, and Minas went away without coffee or money. Soon he came back with the Pungawa.

"Tuna Minas came for his coffee," said the Pungawa.

"Pungawa, since you say so he shall have it."

So they sat down and talked awhile, sociably, and drank some more coffee and laughed over the joke they had made. And Minas went away with his coffee and his profit.

Thus it was that Minas prospered. After a while he had a house that was much larger than his need, and a big motion picture theatre, and a garage full of motor cars, for the steamship company was beginning to drum up tourist trade. Then Vardon came to eke out a living on Minas's farm.

I disapproved of Minas. I told him he was doing his damnedest to ruin the island, first with movies, now with motor cars and tourists. But so long as he didn't sell shirts, I would associate with him.

"We-e-ll," said Minas, "the Baliers won't go to movies, and they don't like autos, and the tourists just come and then they go. I guess the Baliers will be all right."

To the Dutchmen Minas was just so much scum. That didn't worry him. It was even harder to make money from the thrifty Dutchmen than from the Balinese. He even dealt with the bank as little as possible. The Volksbank was not a real bank anyway, but a Government institution with a monopoly on land mortgages. It lent the natives money at 15 per cent, to protect them from usury. One day the cashier suggested that Minas borrow some money, since he had a surplus. And Minas, thinking it over, decided that he could use 10,000 guilders. But when he called for it, the banker, in a sudden gust of caution, had changed his mind. He was sorry, he didn't have the money.

"So-o-o," said Minas, solemnly. He went home and opened his iron box. He went down to see a Chinaman and said, "Give me 5,000 guilders." The Chinaman did so and they shook hands. Soon Minas was back at the bank.

"I am sorry you have not got money," said Minas sympathetically. "Here is 10,000 guilders I have brought for you."

Three months later he went back again.

"Now," said Minas, "you pay me interest. It is too bad you did not have that money for me. For then I would pay you money, but now, you see, you pay me."

Minas was perfectly content to have the Dutchmen think as they pleased of him, and he kept his own thoughts to himself. But I think it gave him great pleasure whenever a subscription was being taken up for a worthy cause, to give exactly as much as the Resident, more than which no one dared. And there was one time when they wanted a benefit performance for volcano sufferers, and he refused to be paid anything for his theatre. "No," he said. "It is for me to give the show, but you will buy all the tickets." So the show made a lot of money, and I think Minas enjoyed thinking about the thrifty ones who had bought more tickets than they could use.

The Resident asked Minas a favour once. Clemenceau, the great French tiger, was touring to Bali. It would not do to entrust him to a native chauffeur. Would Mr. Minas drive the Resident's car while Clemenceau was here?

"We-e-ll," said Minas. "I will drive Mr. Clemenceau, but not in your car, Mr. Resident. In your car I may be Mr. Minas, and in my car I may be just old Minas, your coolie, but it will be *my car*.

"*Suda*. Finished," said Minas. So Minas, in his own car, drove the great Frenchman. A long time later Minas received a telegram signed with a strange French name. In the morning Minas shaved, put on a clean suit of pyjamas, and went down to meet the boat. A distinguished looking gentleman stepped ashore and walked past the crowd of Dutchmen, straight up to Minas.

"*Bonjour*, Monsieur Minas," he said and shook Minas's hand. "You don't know me."

"We-e-ll, I don't know you. But you know me. And you give me your hand. And I take it."

"Clemenceau told me about you," said the Frenchman. "He said to come to you if I came to Bali. He told me just what you looked like."

So Minas got a car with a native boy to drive it. And he showed Bali to this Frenchman also.

Minas was very hospitable. He gave me a feast the night before I sailed. We drank wine, and he gave us free translations of Persian verse. "Dirty? Yes," cried Minas into our ribald mirth. "But in Persian it is goo-oo-d poetry!"

Minas was always ready to talk. Without stimulus he would tell about the girls he had known, about the time he threw the Regent of Gianyar out of his theatre. Give him a few drinks, he would make bawdy Levantine yarns. But give him a full bottle, and he would really talk about Persia.

Persia. Ah, there was a land! He had left it when he was a boy, but he knew all about it. There was a land of cool, rippling waters, dates and pomegranates, veiled maidens beneath a gentle sun, telling stories all the day. There was a land of plenty and of happiness. Roosevelt and I looked at one another. In our eyes were deserts, dust and poverty. Minas went on speaking. In his eyes were milk and honey. Persia. Ah, there was the land! Let no one tell him about Persia. Was he not born there? Did he not know?

Persia, and the little brother and sister there. They had not been born when he had gone away, but now they wrote to him. Sister, she was a doctor in a hospital. A great crowd had come to the railroad station to meet her. Brother, he was in a great bank. Ah, it was good to know of them. Here in Bali, he was just Minas. But there in Persia, where brother and sister were, the name of Minas was an honourable name. It Stood for Something.

Confronted with such naive humility it was hard to keep from being philosophical. Outside in the night was an Eden warm with moonlight, and all was Minas's garden. Let him but call: Rajas and Gustis would meet at his council table. Let him but speak: a yellow man would give him 5,000 guilders for a handshake. And out in West Bali, on his acres, lived the one-time millionaire. Minas was proprietor of paradise, but he did not see it. Far away was Persia, and a sister nursing in a hospital, and a brother adding figures in a bank. Far away were Persia, and childhood, and they were very, very beautiful.

So Minas would talk, and after a while Roosevelt would go to bed. Minas would go on talking, and then I would go to bed. When Minas was tired of talking, he would go home.

Minas's conversation was quite polygamous, but I never saw but one woman around his place. A good looking girl she was, who appeared as though she might be quite sufficient. And Minas seemed quite satisfied, except that he was complaining about his coffee. One afternoon he dropped in to call.

"Have you got coffee?" he asked.

"You know, I had to send that girl away. You know, I found out what was wrong with my coffee. She have been putting *guna-guna* love charm in it. . . . *Suda!* Finished! That son of bitches."

XXII

I WAS in Buleleng. Roosevelt was in the South, and I was to meet him there. There were to be great days, feasting at Tampaksiring, lazy hours at Bengkel. My Armenian friend Boldi, who spoke some English, arranged for me to ride down in a native car, connecting with a bus, for five guilders. That would not be exactly upholding the prestige of the white man, but it would save ten dollars.

The car came for me at ten o'clock. There were two or three natives in it, and a Chinese chauffeur. I put my bag in behind, and took the place of honour beside the driver. We rode about a bit, stopped at various places, though I could not find out why. The places were not far apart, but it required considerable travelling to get to them, for Buleleng has one-way streets. Your driver may go to sleep at the wheel, but otherwise out here they are very careful with their cars. In a one-way empty street, he klaxon shrieks hysterically. As we lingered down in he bazaar street, considering that we might arrive rather ate, I took time to eat some *sati*.

There are two ways of eating *sati*. One is to summon the *sati* man while you are eating dinner. He may squat out front by your doorstep; or, if you are inclined to go in for atmosphere and like your *sati* really hot, you may have him bring his portable restaurant inside, beside your table. He carries his shop in two boxes slung on a pole over his shoulder. He squats between them, fanning charcoal in a narrow tray, fierce charcoal burned from the shell of coconuts. Then on a grill he ranges bits of fat meat, skewered on little spits of wood, like lollypops. Probably it is goat's meat, or chicken, for he is a Mohammedan from Madura, up north of Java, and will not touch pork; while the Balinese will not eat beef. He dashes the bits with coconut oil, fanning them the while. Meanwhile, in a Chinese dish, he is mixing a sauce of mysterious ingredients. Its base is the *lombok,* a red pepper of size and potency comparable to a .32 rifle bullet. There is soya in the sauce also, and other things I do not know. The *sati* is brown and blistering hot, and you dip it in the sauce before you eat it. This manner of eating *sati* is quite compatible with white prestige.

The other way of eating *sati* is somewhat impromptu. Perhaps you have been eating dinner in a Chinese shop, which no white man ever should do. You have had a big dish of chopped cabbage and other green things, well scalded and mixed with shrimps. You have had rice, and a fine big omelet with prawns in it. You have eaten a great deal and wish you had ordered twice as much, but since it would take time to cook more, you go hunting for a *sati* man. Probably you find two, in the street outside; and being fair, you start a competition between them. You tell each to grill six pieces, as a sample, which will make three for each member of the party; and then you sit down on whatever may be handy and wait. Since each of the samples turns out to be equally good, you put the cooks to work in earnest.

The *sati* is very tender, for the meat was killed at daybreak and has by this time lost all animal heat and toughness. It is piquant and indefinably savoury. After the third dozen pieces you wish you hadn't gone to the Chinese shop at all, and after the fourth dozen you give up. Then, unless you object to the *sati* man's prerogative of overcharging a white man, you pay one cent, or two-fifths of an American cent, for each piece of *sati*. Really this is not too much to pay; for, eaten in this way, *sati* is superior to any other food in the world, except the *sukyaki* they give you in Japan.

I say this in full cognizance of the green pea soup at the Savoy-Continental at Cairo and of the sole Normandie at the Restaurant Isnard at Marseilles. For to appreciate fully these latter gustatory poems, one must have spent weeks coming out of the tropics on a ship whose fresh foods have given out, weeks of heaving through a monsoon on the Indian Ocean in a sixty-mile gale of Turkish bath.

And now, if you are wondering what all these comments have to do with this story and wish I would get on to Den Pasar, I have succeeded in putting you in a state of mind somewhat similar to mine on this particular morning. I wanted to get started. I was hot and bored, and couldn't see any reason for not getting started.

"*Musti pigi Badung,*" I exclaimed, by way of saying that we must go to Den Pasar.

"*Nanti, tuan,*" which meant that we would start sooner or later.

"*Musti pigi scarang!*" I insisted on going now.

"*Tida bisa, tuan.*" *Tida bisa,* no can do. There are three scraps of Malay that any Balinese knows, and they mean "can't do," "don't know" and "don't dare." One of the three is the first answer to any question or request. And quite logically; for if accepted, it will save a lot of effort. But if you insist, you may stir up some of that effort.

"*Musti pigi Badung, scarang!*" I cried. But what was the use? I knew we would go after a while. It was too hot to get excited. I just waited. One learns patience in these latitudes. Once in a while an American burst of hustle will crop out, but what's the use? Things will be done sooner or later, and time doesn't matter. One doesn't appreciate the value of this lesson of the oriental tropics until one reaches the pier at Manhattan, and waits to get ashore.

Sure enough, after a while we got started. We had taken on considerable cargo. The stern of the car was lashed heavily, with a nest of iron pots and various other commodities. There were several bags of vegetables in the back seat. There were six natives there also. One of them was dangling over the starboard side. And so we rolled out merrily through the paddy fields. When we came to a slight grade we went into second. It was, mercifully, an uneventful climb. The car was a Dodge, whose paint was still shiny; and I pay it the tribute of recording that by one o'clock we arrived at Kintamani, on the ridge, in low gear.

We stopped awhile by a roadside shop. A passenger got out, and two more got in. We waited several minutes, but I was not moved to protest. It was pleasantly chill up here. An uncouth, dirty looking people, these mountaineers, but sturdy withal. They hike down to the lowlands with great loads of cabbages on their heads. We had a whole family of them on our front porch one night. They had come shivering along the street in a terrific drenching rain, and Roosevelt had called them in. He gave them cigarettes and a sheltered place to sleep.

A good many of the mountain women had great goiters, strangely, for you never see them in the lowlands. That reminded me; I had meant to ask about the goiters, when I talked to Dr. Onkieong, the able Chinese physician in charge of the hospital at Singaraja. He had remarked that there was excellent health in the mountains, no tuberculosis and very little infection by the *gonococcus,* which all tropic doctors must so generally fight. The health of all Bali is good, and its half dozen doctors have done much to improve it, with vaccine and salvarsan. Strange, I thought, that in this island where the *spirochæte pallida* is such a troublesome parasite, I should see less signs of its depredations than one might see in a few hours on a New York street.

We were off again, rolling down through the southern villages. It seemed that the Dutch had done wonders in their road building. Yet later you realized that in these lowlands they had but connected and surfaced the broad avenues of the village streets. Any dry day you could turn off this road and drive for miles through broad avenues paved with grass. There were villages, villages, villages, dotting the countryside away from the roads.

And so we rolled down through highways teeming with human life. Strange, I thought, that in some of these towns the people should be so ugly, with loose, protruding mouths and blobby noses. In Bengkel, and other places I knew, the people were quite generally handsome. Inbreeding had done it, I suppose, centuries of marrying the little girl next door. I used to wonder, sometimes, when I found myself in one of these less favoured communities, how I could have been so ecstatic the day before in Bengkel or Ubud; wonder whether I had been looking through rose-coloured glasses. But in such skeptical moments I had a sure cure for doubt: I just imagined Philadelphia, or any other American city, all garbed in the Balinese fashion.

Thus, thinking of this and noticing that, I came soon to the crossroads near Gianyar, where I was to catch the bus. And just then, roaring along the road from Gianyar, came the bus; and with much waving and shrieking of the horn my chauffeur halted it. I found myself left at the crossroads, with a new vehicle and new companions.

This bus was a weird mechanical apparition. It had been, apparently, a small touring car; but now a vast box, labelled for eighteen passengers, was balanced over its rear axle. On the left side was lashed a cylindrical bamboo basket, and within this was trussed a hefty pig. The box had its full quota of passengers, and bulging up between them were great gunny sacks and bunches of bananas. As I gazed dubiously at its hood, I discerned indisputable signs that the vehicle had been a Dodge. There were battered remnants of what had been acetylene lamps. The driver smilingly invited me to take the seat beside him.

But there were certain things to be done before we started. The assistant chauffeur, who had put a rock under the rear wheel when the bus had stopped, was now unscrewing the radiator cap. He gave it a few turns, then pulled his hand away. The cap shot upward three feet on a column of steam, like a celluloid ball in a fountain. The steam blew off in a tremendous exhalation. Then he was pouring water into the radiator, from an oil can. It blubbed and sputtered, but finally accepted a long draught. I climbed to my seat. Now was the crucial moment. The assistant prepared to crank. Could this thing possibly start? A single twist made a shaking roar. A gallon of water flew out the left side of the hood. Blue smoke fumed up through the cracks by my feet. We were off in a vibratory rattle. I was sitting on what had been a spring cushion. I moved myself about a little, to find a cavity that would fit.

The driver was quite skilful, a shock-headed, barefoot youth with knickerbockers and a striped shirt. The car careened along for five kilometres before it stopped again. The assistant leaped out and put a block beneath the rear wheel, then addressed himself to the radiator cap. Astonishing, the skill he had developed. I marvelled at it all that afternoon. Just the right amount of twist, and then his hand would jerk away from a spurting, scalding geyser that sent the screw-cap flying. Astonishing, too, that radiator; it had three holes as big as my fist.

Flying along the road we went, ten kilometres more, with only one stop for water. That stop was in a market place, and some of the passengers bought pink sugar water to drink. It was in bottles that had wooden stoppers shaped like snipes' heads. The beaks were not perforated, but had grooves down the side. A bottle was tipped up, a head was tilted back, and the sugar water ran down the groove to drip into an open mouth. On we rattled, and at the next watering stop the car halted, sputtering. When it was cranked, there was only silence. The assistant cranked furiously, the sweat poured from him, but from the motor came not a cough. Investigation. Consultation. Discovery. The young Chinese in white pyjamas, who seemed to be the conductor, explained to me. The gasoline was all gone.

There was no consternation among the passengers. They got out and stretched their limbs. It would not be long, said the young Chinese. We chatted for a while. He taught me some Malay words. The passengers sat down by the side of the road. One went on a sociable hunting expedition in the chauffeur's shock of hair. They paid no particular attention to me. It must have been strange **for** a white man to be riding in their bus, but there was no prying, fawning curiosity. The Balinese are a great people for minding their own business. I learned this in some discomfort in my first days in the island. Smiling sociably at little girls, I felt suddenly as though I needed a shave and had inadvertently spoken to a strange debutante on Park Avenue. Anyway, the passengers settled themselves down to wait, seeming to mind the delay not at all. Nor did I care much. We had stopped in a place of coco groves and paddy fields, just after rounding a bend. By the bend stood a tall tree, such as Corot might have painted. Beneath it was a temple with lazy cattle grazing by. I got out my camera and wandered back. I was still taking pictures in those days. I had not yet learned how hopeless I was as a photographer. After a half hour or so the assistant chauffeur came back with a can of gas, and soon we were roaring on down the road. We came to a suspension bridge after a while. Everybody got out, but me and the driver, to walk across. There was a truck by the bridge, loaded with great sacks of rice, and coolies were unloading it. They would carry the rice across the bridge, then would load the truck again when it had crossed. The Dutch were very careful not to overload these bridges.

Along the road we thundered, ten kilometres more, with only one stop meanwhile for water; but as we stopped the fifth time an acrid odour suddenly smote the atmosphere. Smoke! The left hind axle! A hotbox, sure enough. Angrily the grease was bubbling, and sharp blue fumes poured forth. Philosophical, unhurried, the assistant chauffeur filled his bucket at a ditch, and dashed the water into a steaming cloud about the wheel. The driver took his pliers and loosened the brake band.

We were off again. Two kilometres more rolled by. But something was wrong. The road was level, but the car was dragging. The driver went into second, and the speed slackened still more. He shoved his gears into low, and the car stopped. More smoke. Flames, bursting from the left rear wheel! The emergency man got out his pail again, and drenched the fire out. Then it was plain for all to see that the axle had melted off, and that the wheel, jammed up against the fender, still was propping up the car. The chauffeur gazed at it sadly, helplessly.

The passengers all had dismounted. They examined the wheel also, and then set about unloading their luggage. There was not a bit of excitement. Not a voice was raised. They gathered around the driver, paid their fares, and then strode off down the road, with great piles of dunnage on their heads. Their feet and heads had done good service, long before the white man brought his strange contrivances.

I stood there disconsolately, with my bag beside me; and the Chinese youth in white pyjamas gazed at me, perplexed.

But our difficulty was not for long. In a moment down the road came thundering our bus's counterpart. We hailed it with a shout of glee. It was only two or three kilometres to the rest house. It was five o'clock. How I should eat! I shared the front seat with an old white-bearded Mohammedan, with an infant in his lap. He gave me an ugly look as he moved over. I did not blame him.

Three days later I went along that strip of road, and there in its midst still stood my bus, deserted. I liked to see it there. Bali has acquired the motor; but, praise be to Shiva, she has not yet acquired Efficiency!

LIVING TREASURE

XXIII

MY precipitate ride had cast me into a new world. It had been a disenchanting afternoon, of heat and barking dogs and glaring sun that turned the world all black and white. Now as the daylight softened I had no sooner sat down on the rest house verandah with a large whisky soda and a small cheese sandwich before me, and begun to recount the marvellous mechanical events of that afternoon to a tourist I had buttonholed, than Roosevelt appeared.

"The *gong*," he heralded. "The Belaloan *gong* is playing."

"Another drink," I said, "and we will go."

For I knew this Belaloan *gong* well. It was the finest *gamelan* in all Bali. The *banjar* was just a few steps down the road from this Den Pasar rest house. More than once, killing time at midday, I had stopped in there and taken an impromptu music lesson from small boys who were practising. You could hear music there at any time of day or night. Sometimes for three hours every night for weeks the *banjar's* orchestra would practice, a club including the highest caste men in the village and some of the humblest coolies.

Belaloan is the centre of the modern music movement in Bali. If in speaking of this civilization as a survival of ancient times I have given the idea that its life is static, I have been careless. For this life is alive; and in art, dancing, and especially music it is developing and changing all the time, along its own peculiar channels. The classical music is never played twice in precisely the same way. Its form is laid down as an immutable skeleton, but within its narrow limits the composition is dependent upon the inspiration and skill of the individual performers. But in recent years a more stringent discipline has been laid upon the players. They have been working out modern compositions to the finest degree of subtle detail.

Listening to one of these modern pieces, I sat enthralled one night at Belaloan. It was of eight or nine short themes, each repeated a dozen or more times in exquisite variation. The *gamelan* played like an individual virtuoso. Two instruments struck alternate notes in passages too rapid for one man's sticks. Tone diminished to scarcely perceptible trickles; it burst forth in breathless attacks; it built itself into magnificent climaxes. The orchestra played like a single inspired organism.

And all without a leader. This may seem strange to those accustomed to the flying hair and batons of star conductors, but it is the very essence of Balinese character. The *gangsas* start themes, a drummer guides the rhythms, and for various refinements of execution the players depend on a half dozen of their number. Neither did this

music have any individual composer. It was the work of
the whole orchestra, developed from ideas as welcome
from the youngest as from the oldest player—tried out in
rehearsal, modified, accepted, or discarded by common
consent.

But on this evening the Belaloan *gong* was not playing
at its own *banjar*. A Gusti from some leagues away had
dropped in to visit a cousin in one of the *banjars* at the
edge of town, and being an expert of some reputation at
the *topeng* dance, he was giving a show for the neighbours. The generous gentlemen of Belaloan were contributing the music.

The dance was on as we arrived. The whole village was
clustered about a dilapidated temple gate. Dozens of
children perched upon the wall. Dogs were running
around among the crowd. Curs are everywhere in Bali,
barking nuisances; Den Pasar especially has its hosts of
them. But now, as music filled the air, the dogs were silent.

Here was a picture in strange contrast to Bengkel, only
a mile away. In Den Pasar, a thriving town, the ways of
the whites were coming in; and fully half of this celebrating crowd were arrayed in nondescript shirts and coats.
The garments were an obscene incongruity. But underneath was throbbing the true, unblemished life of Bali.

In a little space before the orchestra the Gusti danced.
He was heavily draped in golden *kains* and at his back was
a golden *kris*. Over his face was a *topeng,* mask, of quaint
inhuman ferocity. He wove about with spraddled knees.
His hands and elbows twisted, now sinuously, now in
sudden angles. His rough, guttural voice burst from be-

hind the mask. The children shrieked with mock terror and laughter.

The dancer had a little curtained tent, and from time to time would enter, to emerge in a new mask. He was acting all the characters in a play, and with each mask he changed his voice. His masks were of infinite variety and delicate workmanship. Now he was a ferocious demon, now an aristocrat with elaborate moustachios; now, sure enough, a caricature of just such a comic Dutchman as I saw yesterday.

This dance was interminable and unintelligible to me, but always there was music. Music—this indescribable, untranslatable tongue. Music—the disembodied apparition of the soul of Bali.

In Buleleng life had been a gay, lusty, bohemian white man's thing. Now mystically the world was changed. The gongs thundered, the *gangsas* tinkled, the *joblags* hummed. Night fell and palm trees hung their silhouettes across the sky. Music coursed through the veins, an exquisite *katharsis* of the spirit.

"The Belaloan *gong* has been working for the last month on a new piece," said Roosevelt. "They say that in two months they will play it for me."

XXIV

BUT even at such times man must eat. Madé was waiting at the gate to take our bags when we got to Kumis's house. Babad and a lot of his friends were out there too. Réné cooked dinner for us. She spitted a chicken on two crossed sticks and broiled it over charcoal with coconut oil and a fine red pepper sauce.

We had brought some ice along, and I mixed up a tall Tom Collins, or whatever it is when you use a *jerup pipis* instead of a lemon. It was pleasant there in the friendly night. Our *jerup* drinks made us feel like sun-dried leaves relaxing in the dew.

The chicken was savoury. We had a loaf of bread we had got in Den Pasar. Bread in Bali is a foreign barbarism. This loaf had been cooked in an oil tin. In the centre was a cylinder of half-baked dough that Roosevelt pulled out and threw away. The rest of it tasted good. We ate it with chutney. We would have a good curry tomorrow with that chutney.

I offered Madé a swig of my second Tom Collins. She drank pretty near all of it, and handed back the glass with grave silence. Madé was strangely quiet tonight, even for her, who never talked much. I made a new Tom Collins for myself.

In the cluster of houses at the corner of the compound an old man's pinched voice crooned an incantation. Over a thatched wall dropped gently the notes of a bamboo pipe. From the *balé banjar* came the pounding of a drum.

Madé sat there at the edge of the lamp glow, smoking my cigarettes, one after another. The outlines of her melted in the shadow. When she drew on her cigarette, it lighted up her face and negligent hair, and her great eyes. Then for a while all I could see were occasional curves, a breast, a shoulder, a cheekbone, where the lamplight touched. After a while Madé spoke.

"Ida Bagus Gidéh has got a new wife," said Madé.

"What! Ida Bagus Gidéh with three wives! Who?"

"A Child of the Great. A Satria of Kesiman. Sayu Putu Rapug."

Sayu! The golden girl of the crumbling palace! Ida Bagus Gidéh had married Sayu!

Now at once everyone was telling us the news.

"Ida Bagus Gidéh's wives are angry," said Réné. "They say Sayu is no good for working. They say all she can do is comb her grandmother's hair. They say she will just make more work for them."

"Ida Bagus Gidéh's father says Sayu cost too much," contributed Kumis. "Ida Bagus Gidéh gave a hundred *ringits*. The Padanda wants him to send Sayu home."

"Ida Bagus Gidéh had to get his brother to talk to Sayu's father, after they ran away," said Madé. "The Padanda wouldn't do it."

"The Padanda says that those Kesiman people are just upstarts anyway," went on Kumis. "Sayu's grandfather was just a Pungawa and he got an army and called himself a Raja. The Padanda says that anyway the Satria are too proud, and sometimes they even claim they are higher than the Brahmana."

"But Sayu has beauty. Sayu has friendliness," suggested Roosevelt.

"*Ada, tuan,*" everyone agreed.

"Ida Bagus Gidéh and Sayu are at the *janger* school tonight," Babad informed us.

Thus piped the voices of the *janger* girls. I don't know the words. They meant nothing anyway.

Ida Bagus Gidéh was sitting there in a *balé* at the side, arms weaving as he taught a lesson in the *gandrung* pose. Sayu sat beside him. In a *balé* beyond, grim and sombre, I could see the faces of his other wives, the plump and scrawny ones.

What a woman Sayu was. As she sat there cross-legged, silent and serene, there was no particular mark that made her more than one woman among many. There was nothing that a camera could catch, save the subtle modelling of her brow and breast. But there she was, and the whole gathering at the *janger* school seemed but a setting for her subtle emission of glamour. She was a princess, that was all; and the lift of her chin, the slow turning of her head, announced it.

Ida Bagus Gidéh caught our glance. His face turned from a dancing mask to a friendly grin, and with swift sinuous tread he came toward us with hand outstretched.

"*Slamat*. Greetings and happiness to the bridegroom."

"Thank you, *tuan*. I am very happy. But my family is very cold to me."

"You have got a very beautiful woman."

"Yes, but my family want me to send her back. I will not do that. My family are cold to me. But Sayu is warm."

Sayu arose with placid dignity as we approached, greeted us with a slow luminous smile. We stood there chatting, but she hardly spoke a word. I do not remember a word she ever said. She was one of those women who do not need to talk.

As we stood there the glamour of her spread like slow stimulant through the veins. I coveted my neighbour's wife.

Sayu smiled, and my eyes were dazzled with the wonder of her teeth, those large, white, even stones unsmirched by betel. Ida Bagus Gidéh grinned happily, and my breath caught with a sudden sense of foreboding. Ida Bagus Gidéh's teeth had been filed. Sayu now was married. Sayu's teeth . . .

Throughout these islands of the East there is this strange, vandal custom. In the jungles of Sumatra tribesmen horribly mutilate their jaws. The Moros of the

Philippines cut their incisors almost to the gum line. Bali
in the old days reckoned its population by the number
of men whose teeth had been filed at puberty, who now
were ready for war. And the filing was a test that brides
in marriage faced. In Bali the custom has been dying out,
but to many of the aged has come its grim and horrid
reckoning.

Ida Bagus Gidéh's teeth had been filed. Not enough
to disfigure him. But their edges had been cut to a trim
and even line.

Never had I been able to get any reasonable explanation for this custom. I would try again.

"Ida Bagus Gidéh, why do you file your teeth?"

"Because," said Ida Bagus Gidéh, and he paused with
a puzzled frown. "Why, *tuan,* if a man's teeth weren't
filed he would be just like a coolie."

"And Sayu? Will her teeth be filed, now she is married?"

Ida Bagus Gidéh grinned, then looked very serious.

"No, *tuan.* You know, the Chinese man wanted two
hundred guilders to file her teeth. That was too much."

Ida Bagus Gidéh looked very solemn. He glanced at
Sayu. I think there was a twinkle in his eye.

After a while Ida Bagus Gidéh and Sayu went home.
What a handsome couple they were as they walked away,
with their sloping shoulders and deep-shadowed backs.

How I envied him. He was a prince in paradise. For
him moved this gentle, shining stream of life in Bengkel.

For him the moon shone on the palms, for him the night was amorous. For him there would be dancing and singing, and the *gong* would play, while eternally bloomed the frangipani and hibiscus flowers. He was a prince in paradise, and he had found his princess.

The *janger* sang its lilting dancing song, and hands flew gaily in the lamplight. Now beyond the dancing square there passed a pair of quiet-moving figures. The wives of Ida Bagus Gidéh, the fat one and the scrawny one. They passed the lamp and threw long shadows on the joyous scene. They passed on into the night, a pair of silent doubts.

Ida Bagus Gidéh, was he but human after all? Was he like us all a vagrant in pursuit of dreams? And did they fade at touching?

That plump wife: four years ago was she the vision of heart's desire? The thin one: was she once a fragrant wisp of loveliness? And Sayu? . . .

The *janger* shouted, rollicked in its dance. The one-string fiddle whined.

Now, from the row of kneeling dancing girls arose a tiny slender form. Renang! I had forgotten her. She wheeled in the centre of the dancing square. Her arms extended were a wood nymph's incantation.

Renang was growing up. Renang would soon be ripe for marriage. Beneath the binding of her golden scarf shone promissory undulations of her budding breasts.

XXV

HAD there not been a chunk of ice left, wrapped up in a blanket, we should have gone to bed. It would not do to waste the ice. So we had nightcaps.

"You speak of the quiet excitement of these village nights," said Roosevelt, quaffing his whisky soda. "Save your excitement for the feast of Tampaksiring. Or for a big cremation. It's a shame you aren't going to be here for a good one.

"What! You think that cremation you saw the other day was a big display? It was nothing. Only nine bodies. Hmph.

"They ought to be having one right here in Bengkel pretty soon. Kumis's father died two years ago, but I don't think they have burned him yet. When that time comes, I tell you, Kumis will be digging up not only his father's bones but a lot of his hoarded *ringits*."

Roosevelt turned to Kumis, who was sitting there near us sociably with Madé, though they couldn't understand a word we said. When were they going to have their cremation?

"Nanti," said Kumis, "later."

"Well," went on Roosevelt, "the Raja of Bangli died two years ago, and that was a time. Let me tell you about it. A really good funeral doesn't happen very often.

"It was a sad occasion when the old Raja died, of course. He had been ruler for many years, and the people loved him. But still the event had its compensations. For three years all the people who died had been accumulating in the burial grounds, just waiting for the death of someone of sufficient importance to warrant a festival. They all could be cremated now.

"There was old Wyan, for instance, the father of that boy who works for Minas. He died just after I came to Bali.

"When Wyan died his body was laid out in state for a while. A gold ring set with one of those pink rubies was placed on his tongue. A piece of iron was fixed between his teeth, and a melati flower between the molars. A tuberose was put in each nostril, wax in the ears, and a little mirror on each eye. All these assured that when Wyan comes back to earth he will have good teeth and an excellent assortment of senses.

"There wasn't much ceremony about Wyan's burial. They took the ring out of his mouth and replaced it with a red flower, then put him in the ground. The *gong* played, some offerings were given, and some incense was burned. He had no coffin, but there was a hollow bamboo which was placed upon his chest, extending up to the surface of the ground.

"This tube let Wyan's soul out. During the years before the Raja's death a great many souls had accumulated,

which were wandering around the neighbourhood, though
never very far from the bodies they had inhabited. Now
they could have a cremation, and it would be a joyous
occasion, for it would release all those souls and let them
go to heaven. It would clear the atmosphere.

"So now the old Raja was dead. He was not buried. A
special building at the palace was set aside for him. He
was wrapped in a shroud of white cloth, a hundred yards
long, with all manner of spices. He was laid out in state.
Then all the priests gathered, consulted the moon and
stars and ancient books, and set a date for the cremation,
two months ahead.

"There was a lot to be done in those two months. Suddenly the people were working with bamboo, and *papier
maché,* and paint and paper. As the days went on the
activities of the town became more and more feverish.
Old Wyan's remains were dug up and laid out in a *balé*
at his home, wrapped up in tissue paper. There wasn't
much of him left. In dozens of other homes they had done
the same thing. When all the work was done it was nearly
time for the cremation. They took the Raja's body and
laid it in the Temple of the Dead. There were fifty other
bodies laid in state that day.

"Then the whole town cut loose to have a good time.
You have heard the *gong* play, but you haven't heard six
gamelans all at once. This beat any boiler-works you ever
heard. I wish Stravinsky had been there; he would have
gone and soaked his head in a bucket of water. It wasn't
just the people of Bangli now. Throngs had come from
all that part of the country. Their *gamelans* had come

and their dancers, too. Everywhere there was a *gamelan* playing, and everywhere you looked there were dancers and crowds watching them. The *legong,* the *topeng,* the *janger,* the *gandrung,* and all kinds of dance dramas. Everywhere were masks, ludicrous and grotesque. We used to call it dancing in the streets on the 14th of July in Paris; but this, I mean, was dancing in the streets! In the night they had the *arja* and the shadow plays. It went on from dawn to dawn. Nobody got any sleep.

"The *balé banjars* were a vast hullabaloo of cooking and feasting. They were roasting pigs, on spits over coconut husk fires, and broiling chickens and ducks, boiling rice, chopping coconuts and peppers. There was food for everybody and everybody was eating. There was plenty of palm wine, too, and *arak,* which has more power. But nobody got drunk.

"In and out of the temples poured women with offerings on their heads, women in *kains* woven with gold and silver, and many carried golden platters. The priests were there in all their regalia, sitting in their little bamboo pulpits, ringing their little bells, burning incense. The town was festooned with palm leaf offerings, in all manner of intricate patterns. Great bamboos curved into the sky and fluttered palm leaf tassels.

"You know how teeming and crowded any town here is at normal times. Well here was Bangli with the people from all that part of the country come to enjoy themselves. Don't mistake. Here was none of that roistering boisterousness you would find at a festival of superior Nordic blonds. In the midst of all this surging and excitement and swirling of crowds, you could pick out any single person at any time; and unless he were smiling and joking with a genial neighbour, he would be just as calm, cool, and peaceful as if he were sitting somewhere in the shade, meditating.

"On the morning of the third day everybody got up early. This was the big day. All of the Sudra men went down at daybreak to bathe in the creek and purify themselves. They just wore loin cloths this day. Some of the more modern ones had on those little short pants that they wear beneath their *kains*. They bathed and then with a shout came running up to town for the work they had to do.

"The town was aswirl. If this had been New York there would have been 2,000 traffic cops out to handle that crowd. There would have been a Grand Marshal prancing up and down on his horse, and a lot of pompous citizens wearing badges labelled Committee. Here apparently nobody was in charge. There was a lot of shouting and rushing around. Nobody seemed to be directing affairs, but somehow things got done. The only person to get very much excited was the Dutch *Controleur*. He was running around in a great way, grinding his teeth. This was a State affair, so the Government had gone over the plans carefully with the new Raja and his officials and had fixed up a regular schedule of events for all that day. Now nothing was going according to schedule, and the *Controleur* was all worked up. Nothing did go according to schedule, but everything was done.

"The sun roared up over the Great Mountain, and the

procession was ready to begin. How can I tell you about it? How can I put it into words?"

Roosevelt quaffed a long drink and stood up. He threw his hands above his head, and a chicken sleeping on a rafter fluttered and clucked. Roosevelt drew his hands down over the thin hair of his head. Suddenly he sat down and turned to me, his face alight with excitement. The words poured from him, and his hands flew about him as he talked.

"You are standing with me at the foot of the little hill by the cremation ground. Opposite arise the tall gates of the temple, with demons glaring down. People are swarming all around, and sitting on the walls beside the demons. Over the road arch tall bamboos, with their fluttering palm leaves.

"From over the hill comes music, the sound of a *gamelan*. But it seems strangely unlike the *gamelan*, for its harmony is in a joyous major key. It is the *anklong*, and over the hill its men come walking, carrying their bells.

"Behind them appears a vast masklike apparition of red, blue and gold. A dragon's head, the great glaring face of the *naga*, the sacred serpent. Then its great cloth winding body, carried on the shoulders of three dozen men and women. They are the royal family, and the household servants. The women's *kains* are red and green and yellow; gold ornaments gleam in their hair. *Kains* gleam with gold leaf, and gold *kris* handles shine at the backs of men.

"The Raja's widows follow them. There are not many. Most of them have died before. These women do not

look very sad. Not many years ago this would have been
their own funeral, and they would have burned upon the
pyre. They haven't anything to complain about.

"But now beyond the ridge have appeared the cloud-
touching towers of a distant gleaming city. It is moving
toward us. Over the hilltop comes a vision such as you
have never seen. It is the Raja's pylon, a soaring pinnacle
of white, and blue and gold. It is borne on the shoulders
of half a hundred men.

"Its base is wide, with door all carved and painted like
a temple's, and above the door is a great fierce mask
flanked by vast spreading wings that curve upward twenty
feet, all filigreed in gold and silver. Between the wing tips
hang golden *kains,* within which lie the remains of the
Raja. Above this shoots upward on bamboo poles the
shaft of an elongated pagoda. Above the shoulders of its
bearers the pylon soars forty-five feet high. Behind it,
moving down the road, five others pierce the sky.

"See those coolies carrying it, edging it down the slope.
The air is tormented with the clash of *gamelans,* with
shouts and snorts and labored breathing. The coolies'
magnificent muscles swell, knot and ripple under their tremendous load. The sweat pours from them and their wet
limbs are burnished in the sunlight.

"Behind each pylon moves an apparition, a great beast
which will be the coffin for the burning. The bodies of the
Brahmana now are passing, the priestly caste of the phallic Shiva, and their symbols are the sacred cattle. Cows
for the women, bulls for the men.

"There is no mistaking the sex of these beasts. There

is a bull, with great, curved, spreading horns. His eyes are red and his nostrils flaring. And there beneath his mountainous flanks extends the worshipped phallus, the organ of generation, large and artfully modelled, distended and red as with blood. The pulling of a hidden string animates it to living spasms.

"Coming also are the coffins of the lower castes, a vast menagerie, a comic Apocalypse, a nightmare of Noah. Donkeys, goats, a lowly fish, and all manner of strange and monstrous beasts that man has ever dreamed of.

"And always the smoothly moving single files of women. Their breasts are bound and their right hands, upraised, balance offerings on their heads. They bear pottery dishes, platters on stems like goblets, modelled of rough red clay.

"This is but the prologue.

"Suddenly the great flat cremation ground becomes a maelstrom of humanity and synthetic beasts. The *anklong* players leap high to the bamboo foundations of the Raja's pylon. Its sweating, shouting bearers break into a breakneck run. The tower sways, careens. Now all the towers and all the beasts are sprinting in a whirling circle around the field of death. The air is raucous with shouts and smoky with dust. Three times around the field the wild procession rages. Go hence, ye evil spirits! Get you gone from this field of sacrament! And surely there are no evil spirits now. No obscene shades could face this exorcism.

"The Raja's pylon is rested upon the ground before a sloping runway built of bamboo, and ranged about the field beyond, before their waiting piles of kindling wood, are all the coffins of the meaner folk. And now that there has been a moment's lull, prepares the climax of this funeral day. Travellers have mistaken this imminent outburst. It is no breaking forth of necrophile savagery. Savage its origins may be, as may be some portions of the Christian liturgy; but it has become a pure symbolic ritual. The runway leads up to the niche wherein the Raja's body lies. Now it is overrun; it sags and creaks with a swarm of straining men who have rushed upward like a tide.

"Out from the niche is dragged the Raja's winding sheet, down over heads it passes from hand to hand, until a runner grasps it, speeds it off through the crowd. It winds, writhes, squirms like an endless white serpent.

"With it, over the head of the fighting expectant throng, comes the Raja encased in matting. A host of naked brown limbs are compacted in a seething mass, as the Raja's followers battle for his body. Over the crowd he goes, a macabre pushball, from pushing hands to pulling hands and back again. The shouting is a roar. Bare feet shuffle the dust to an obscuring fog, and rivers of sweat congeal as mud. Forward and back the body moves, over a sea of hands, hands, hands, stretching, grasping, clutching hands, a choppy fingered ocean.

" 'Take our Raja!' cry three hundred throats. 'Burn him! Free his soul!'

" 'No! No!' wail the hundreds more. 'Do not take away our Raja.'

"These calm-eyed men of yesterday are a mad and stormy surf of flesh. They fight like demons. But slowly, inexorably the body moves, over the vast throng and the sea of hands, toward its pyre.

"A torch flares out. From four dozen pyres smoke pillars arise. The air is rank and acrid with vapours, and over all hangs a dark and sullen pall. The *gamelans* ring, and through the cloud the flames of the six great pylons shoot up like rockets to the sky.

"There are the women also. A laughing woman with a great clay jar of holy water sways like a bacchante dispensing her libations.

"And there close by us, kneeling on the ground, six women give their last word to the dead. They put their clay platters before them. They place there patterns of palm leaf, and bits of leaf and paper whereon is writing in their wormlike script. They pour holy water. They put their fingers tip to tip and bow in prayer. They break their pottery, and then they file away.

"The flames die down. The ashes are gathered up. With a great rush and flurry they are carried to the stream near by. Souls have gone to the ocean of eternity, the ashes are washed down to the sea.

"On the cremation ground all that remain are wood ashes, charred bits of weird beasts, torn fragments of palm and paper, and broken bits of clay."

I poured Roosevelt another drink. There wasn't any ice left but we didn't need it now.

"What's that?" said Roosevelt. "You want to know how much all this cost, you materialistic American? Can you measure souls and death in dollars?

"How much it cost is hard to say. You speak of a dollar, but what does it mean?

"The man of Bangli may work all day for a guilder. Even the wealthy ones are very thrifty. Chickens, at four for a guilder, are luxuries to eat. Five guilders is a lot to pay for a *kain*, which is a suit of clothes.

"But when cremation times comes, there is no pause in spending money. There is no telling how much the family of old Wyan and all the other families spent on that great celebration. But it was part of the *Controleur's* job to figure up the Raja's family bill.

"That bill was 60,000 guilders, or $24,000 Gold."

XXVI

THE last dying drizzle of the rainy season fell as we rode up toward Tampaksiring. Through the mist the afternoon was leaden dark. And now, while we yet were fifteen miles away, the village of Tampaksiring came toward us, met us, passed beyond through the murk, down the road toward Gianyar.

A clang of gongs foretold the coming. Around a bend the villagers appeared, a strange, long procession—men, women, children, bright spots of sodden finery and glistening wet limbs. High above their heads towered tall white parasols that cared not for rain or shine, and were no use for either. Offerings of fruit crowned women's heads. On poles between men's shoulders swung gilded sedan chairs and ornate boxes, wherein dwelt disembodied deities.

The men were singing. From their throats poured a deep unrhythmed monotone that diminished with failing breath, and then was caught and carried on by other voices —an endless, mystic, tuneless incantation. They passed with steady even tread and soon were gone, save for the distant muffled clangour of a gong. And too remained a humming in the ears, a steady, crooning, mystic unison, that may have been but memory.

No light shone in Tampaksiring town when we drove through at dusk, and not a person stirred. We drove under hundreds of great bamboo trunks, festooned with sacred palm leaves, arched over the road that led up toward the temple.

Tonight all was quiet, while thirty kilometres away, in the stream at Gianyar, the villagers washed their gods. But on the morrow, with deities purified, with human souls refreshed by annual sacrament, all would return to glut the gods with feasting.

A solid white uniformed figure loomed on the verandah as we came up to the hill-top bungalow, plainly a *Controleur*. It was to us a depressing sight. Roosevelt and I looked at each other dolefully, and a word with the apologetic *mandur* of the rest house confirmed our thought. Yes, we had engaged our room some days ago, but unfortunately the Government had dropped in tonight to occupy it. Here for the feast also.

Well, there were rooms for the lady travellers we had with us. Roosevelt and I could manage. We could sleep on the porch, or some such place. Anyway, we couldn't drive on fifty kilometres to another bungalow, as we had done last week in somewhat similar circumstances. It was a Government house, and the Government had prior and indisputable right to it.

After eating we sat out for a while talking to the Govern-

ment. He was hardly more than a boy. He was thick-
necked and somewhat lumpish for his age. He gazed
moodily into the current whisky soda. He had his wife
with him, a fluffy blonde girl, bit of Dresden.

How bored she was. Think of it! The only white couple
in their town! Nothing to do. And always the noise of
that crazy music the natives made.

"Noise," said the *Controleur,* his eyes still fixed on the
amber of his tall glass. "Noise."

How much better it was in Sourabaya, went on the girl.
There were cabarets there, with shows and music and
dancing. You could have a good time there. But here, how
stupid it all was. How dull and endless this native dancing, how silly the carving the natives did. Not pretty at
all.

Well, there would be a break in the monotony tomorrow, at this native feast. The other officials would be
here with their wives. That would be nice.

I pitied her, and her husband too. As I lay that night
on the verandah's rattan settee, I could see their night
lamp shining through the cracks of the tightly locked shutters that sealed them in. I was enjoying myself, there
under my single thin blanket in the chill fresh air. Only
one who has spent weeks in the equatorial lowlands can
know the pleasure of being cold. I pitied them for their
odd fate. I was having a good time. But she belonged
in Sourabaya, dancing in a cabaret, and he also, watching a vaudeville dance. I was where I wanted to be. But
they—what a dull time a Mennonite would have in
Paris!

I awoke at daybreak. Before the bungalow was a broad level space, which fell away precipitously into a cavernous ravine; and far beyond it the walls of the valley rose high and steep, in tall narrow terraces. The horizon's black palmy fringe lifted high in copper etching, while down through the chasm trickled murky blue dregs of night. Over the hill beside me came a little road, which dove down the precipice in quadruple zigzag. A small car might descend without too great peril, to wide flats wherein a brook meandered. I could see the stream's outlines now in the dim light. In days of exuberant artistry it had slashed its banks with broad, bold, irregular strokes, but now in the waning of the freshet time was shrivelled into the safe path of dry academic senility.

Here I had seen my first sunrise in Bali. Through the gloom of that first breaking day I could see the great sprawled rectangles of a temple by the stream, beneath a shadowy banyan. Dim shapes were moving there, and as the light strengthened I could see that in the temple girls were bathing. Between stone walls in the midst of the holy place a great fountain welled up, and through twelve stone jets burst glistening from its bonds. It was sacred water, crystal pure. The women bathed and filled their water jars and filed away; and always down from the terrace paths still more were coming, shadowed shapes against the hillsides.

But today the place was changed. The sun flamed up, painted the heaven cobalt, and all the world beneath was green and earthen red. The hitherto neglected temple now glistened with fine new thatch on all the *balés*. New bamboo houses were built outside, and fragile altar places; and there was a wide dancing space with high grass roof and European chairs ranged all around. The two stone demons by the temple gates had been draped with cloth *kains* in festal modesty. Over all was an expectant but deserted air. Two women were in the bathing place. A lonely pumanku, robed in white, waded into a pool and there upon a rock placed an offering of fruit and flowers.

Directly beneath me, along the foot of the cliff, was a row of new shanties. A man in a fez appeared there, and then it was all alive with bearded men in long shirt-tails. Bombay traders, out for business. For five days there would be feasting in this Hindu holy place, and it would be good for Moslem trade.

A *gamelan* clangs behind the hill and the sun touches the zenith. The listless morning bursts into vast exciting pageantry. Yesterday's procession, multiplied and brightened with fresh colour, moves over the hill and down the zigzag road. The air is burnished with the *anklong's* music. It sparkles with gold.

Oh, for a hundred eyes! I leap from terrace to terrace like a goat, swinging my camera, snapping photographs. Down over the zigzag trail the pageant writhes like a great bespangled serpent. Towering parasols. Dozens of shouldered shrines. Girls in *kains* of apple green, women in gold-embroidered red. Offerings upon heads, golden platters, piles of fruit. Men with blue striped spears. The processional *naga* writhes to the temple waters and goes through. The gods are washed anew and worshippers one and all surrender to the purifying fountains. With my camera I leap atop a wall. A Chinese tourist guide is there. He chortles:

"I got a picture of a naked lady! A picture of a lady, *all* naked!"

Clothes are forgotten. Girls, who in the morning chill wear more clothes for bathing than at any other time, now in the pious rapture of the moment give themselves unclothed and unashamed to the pouring crystal of the sacred fountain. Like amber nymphs they sparkle in the waters.

Snap the camera shut! To hell with photographs and naked ladies. No film can catch the glamour of this moment. No gelatine can tell the music of the *gamelan*, the bursts of colour, the swirling tide of human piety. I give myself to the surge of life. Here is the end and aim of living, the fierce identity with something rich and strange, beyond all human ken or feeling.

Under a grass-roofed shed the *gong* of Bedulu is playing. I drink the sound. It pours like drops of crystal patterned in a cloth of beads. Men play as though entranced, their fingers flying in complex variation on a four-bar theme. Vacantly open-mouthed their faces hang, but this no longer is a mystery. Now I know this music. Now it is a part of me. Now always shall I hunger for it. These are not men just playing bells. They are a single organism, merged in something greater than themselves. They are not here. They are somewhere else, somewhere beyond all human qualities, beyond desire, love, hope, hatred, passion, beyond all emotion in some vast unhinted Nirvana of sound, where all reality is the singing of the spheres. Here is this fleshless ecstasy, the end and aim of life, here is this distillate of being, this bloodless merging with the eternal Brahm.

The Regent of Gianyar drives up by the lower road in his great gold-encrusted car. He has built houses here for the five-day feast, and installed his wives. In this moment more than ever, he is a gross and hearty man, a gaudy mound of dark carnality. His shirt is purple and his *kain* a raucous green. This is the man Minas once threw out of his picture theatre because he thriftily crashed the gate. "You may be a Raja and I am only old Minas; but I am a white man and so you are just my coolie." But today the Regent is a Raja, and host to all. Up in the rest house his servants prepare a great *rijst-tafel,* and great will be his displeasure if we do not share his feast. With pompous stride he comes to the dancing space outside the temple, where ranged about on undertaker's chairs are solid Dutchmen garbed in high collars and gold braid.

He sits with crossed knees and underneath his green *kain* shows a violent yellow. Before him kowtows a manservant with fingernails two inches long.

Manservants pass whisky and soda, cigarettes. Down at the end of the dancing space, facing the side and a throng of brown-skinned watchers, two tiny dancers pace the *legong*. The white men can hardly see them, but that does not matter; for they are drinking and chatting, and they do not look. The *legong* finished, now comes a *topeng* dancer before the Raja's seat. With masks grotesque and hideous he carries on his pantomime. In small bored groups the Dutchmen drift away, up the hill toward the rest house.

Inside the temple all is life aswirl. Fruit and palm leaves pile high the shrines. Before an altar a woman plucks a duck. Men and women kneel, put flower petals on their brows. Processions are everywhere. Five women come with offerings in the lowland manner, pillars of brilliant fruit and leaves built six feet above their heads.

Perched high on stilts before the temple gate, sits a Padanda addressing the All High, mumbling the secret veda formulae. Above the priest's ascetic face towers a red crown, bound in gold and brilliants as a seven-fold pagoda, surmounted by a crystal ball. A gold-embroidered stole hangs from his shoulders, and strings of beads are crossed upon his chest. With a flower-petalled wand he sprinkles holy water. His long-nailed fingers weave fantastic patterns. Furiously he rings his little brazen bell. Common men may talk to Shiva, but he only can address Tintiya, the Ultimate, the One Above All, whose only function is to exist.

I climb to the cliff before the rest house, and before me spreads the eddying tide of human colour. Far away the terraces of this mammoth stadium are fringed with human watchers like myself. Before the temple gate two dozen men brandish spears in war-like dance.

Suddenly, as by a signal, two hundred half-clad men swarm and converge upon a bamboo altar house piled high with palm leaf offerings. Furiously they tear at its underpinnings. They pyramid upon shoulders, dragging at the eaves. Men leap to the ridge pole, add their swayings to the labour of destruction. The house is toughly built; swaying, its joints hold fast. Then in an instant all collapses. In another instant all is gone. Two hundred men are homeward bound with relics, like good Christians on Palm Sunday.

Forth from his house steps the Regent of Gianyar, from this distance a splendid figure. Behind him in double rank file four wives and twenty concubines. Their breasts are bound in apple green. Their *kains* are black. The *kains* hang long in front and drag as trains between the women's feet. Their linings splash the red earth with brilliant saffron. Before the Padanda the Regent and his women kneel. Five *pumancus* ply them with flower petals, holy water. Royalty are at their devotions.

Before the temple eighteen green-clad girls sway in the rhythm of the dance. Across the valley, down the red vertical precipice, curls a slow procession of brown women. The temple ferments down below me. Festooned bamboos reach for the sky.

And over all hangs the enchantment of the *gamelan*. Its timeless magic throbs in the air. Is this today or a millenium ago? Or is it the Millenium?

Ring, gong! Perish, doubts! Burn, all-permeant Ether, with thy billion eyes! Sing, all ye incoherent planets, universes! This is the end and aim of living.

Ring, gong! Sing faith, oh *gamelan!* Shout belief, ye Gothic spires! Sing Parthenon and Mosque of Hammid, and all ye works of the poets among men! Burn, ecstasy, thy gem-like flame! I believe in beauty!

"Oh, boy! I'm lucky. Oh, boy! I'm lucky. This is my lucky day."

A phonograph is playing. Lucky, isn't it, that those who bear the white man's burden may sit here on the rest house terrace and hardly be at all aware of the doings of those heathen niggers down below? Hail Thomas Edison, king and disseminator of civilization!

"Rockabye my baby," sings the phonograph. "From a jail came the wail of a downhearted frail."

The *Controleur* is dancing with his ash-blonde girl.

XXVII

"THOSE twenty-four wives the Regent had with him were only the better looking ones," said the Dutchman. "The other twenty were left at home, presumably to take care of the children."

We were eating *rijst-tafel* on the rest house porch. Far down at the other end of the table sat the Regent, a magnificent mountain of purple. A proud and arrogant man, with heavy jowls. How out of tune he was with all the Bali that I knew, the gentle life of villagers! How much in contrast to that simple little man sitting beside him, I. Gusti Bagus Jlantik, Raja of Karang Asem.

I had sat with the Raja the other night before his gold-encrusted doors at Karang Asem while dancers of the *arja* postured in his courtyard. No court dancers, but the little girls of the village. And all the villagers had gathered there toward midnight, to watch the drama that would last till nearly daybreak. There were no long-nailed servants abasing themselves before this Raja. He was a simple man among his people. He had sat there beside me barefooted, in a batik *sarong*. I had liked him. He had been wearing a yankee-style shirt without a necktie, and he had given me frightful synthetic white man's "limon-ade" to drink, but I had liked him. And his goblets were delicately chased with gold.

But today, as he sat there beside the purple-shirted Regent, the gentle little man wore the olive-drab uniform of a Queen's *Stadtholder*.

"The Rajas of Karang Asem held virtually complete sovereignty until 1925," the Dutchman was saying. "They were friendly to the Dutch. And even now the Raja holds title as a peer."

Friendly to the Dutch! Compromise. Why did people, and questions, always have to have two sides? A half hour before, I had given myself with a whole heart to the wordless wonders of the *gamelan*. But now in this real world of words and hard facts, it was somehow different. Nothing was black or white. There was the Regent, whom I disliked. But he would not compromise. There at this minute, in all his gaudiness, he was lording it over his white superiors. And there was the Raja, whom I liked and wanted to admire. He, whose people had been a free nation, sat there now humbly in a uniform of the Queen.

I looked down between two long rows of solid Dutch-men and their buxom wives, digging comfortably at their great mounds of rice and meat. How I hated them. Not with any individual dislike: we had not mingled; I did not know them. But for what they were: the practical ad-vance agents of that great modern Juggernaut that would kill Bali. And yet I admired them. I remembered Java, spawning human life with its cultivated richness. Volcanic garden, teeming with rice and tea and human cattle. Its millions were coolies, true, working for perhaps $.20 a day, but with bovine contentment, making out very well withal. Their very existence, at any rate, was due to these practical white men, who had made the land produce food and men. Not only coolies: there were those long streets of excellent city houses that were the homes of prosperous Javanese. I had to admire these Dutchmen—the world's greatest colonizers, surely. I had seen our own altruistic bungling in the Philippines; these practical *boers* were better.

"Compromise? Yes, Gusti Bagus Jlantik is a compromiser," said the Dutchman. "He never would have been Raja of Karang Asem if he hadn't been. He wasn't the direct heir. There were troublesome men in that family. The old Raja's brother, a fierce old fellow with a bristling beard, is in exile in Java. And the direct heir, the present Raja's cousin, was turbulent too. He lives in seclusion, in a town off in West Bali. As a matter of fact, the soveignty wouldn't have lasted as long as it did, had Karang Asem not been friendly to the Dutch.

"Karang Asem once was very powerful, and from its bluffs on the coast it ruled over part of Lombok, the next island, where there is a Balinese kingdom also. Then there were revolts, and the sovereignty switched back and forth. Now both were ruled by a Raja in Lombok, and now by the Raja of Karang Asem. When the Dutch came, Lombok was the ruler; and in time of war great signal fires would flare out on the heights of Lombok. Then the men of Karang Asem would sail out in their *prahus* across the strait, to aid the war.

"In Lombok the Balinese ruled over a great many Sasaks, Mohammedans, who were very much oppressed. The Sasaks complained often to the Governor in Java against their cruel lot, and really such oppression could not be permitted to continue. Expedition after expedition was despatched to Lombok, to free the Sasaks, without much success; and finally there came a great disaster.

"Some Russian soldier of fortune had come to Lombok and cast his lot with the Balinese. (Troublesome fellows, those Russians; some of them lately have stirred up Communism in Java.) Anyway, the Dutch forces at that time were to be equipped for the first time with repeating rifles, but the cargo of arms was shipwrecked, and the expedition went ahead without them. Meanwhile (whether from the same cargo or from another source, I don't know) the Russian had equipped and drilled the Lombok forces with repeating guns. The expedition came, and the signal fires burned on Lombok, and the Pungawa of Karang Asem sailed with his reinforcements to swell the defending army."

My companion paused. He had put away a large stack of *rijst-tafel*. Meditatively he broke and nibbled a wafer made from fish flour.

"It was frightful," he went on. "Merciless slaughter. The Queen's expedition was quite wiped out. The repeating rifles did it.

"Of course," he went on, "that had to be avenged. When next an expedition went to Lombok, the Pungawa of Karang Asem stood on his bluff and watched the signal fires across the strait. But his *prahus* did not put out from shore. Lombok was conquered. The Pungawa became again, and remained, the Raja of Karang Asem."

"Wars," I said. "I'd heard about the old wars, but never really until today did I realize that the Balinese once were fighting men. Never until I saw those men dancing, brandishing spears, did it really dawn on me. It somehow seems inconsistent with the gentleness and peace that is everywhere."

"Oh, they were fighting men," said the Dutchman. "Hand-to-hand fighters. And you mustn't be confused by the gentle simplicity of the village folk. They always have been under authority. If you had mingled more with the Rajas, and Gustis, and Jokordas, the ruling class, you would realize how vehement they could be. The traditional warrior is of the Satria caste. But all of them could fight.

"Now, that fellow over in West Bali, who claims to be the real Raja of Karang Asem. He's one of these men who won't yield. He causes trouble.

"Eight or nine years ago the widow of the old Raja died. The old lady was laid out in state in a *balé* at the palace. Of course her cremation was to be a great event. But then it developed that she could not be cremated without the presence of the pretender, who was in exile. So the body just lay there. I think it was embalmed in some way. They used to practice in the old days a way of preserving bodies with benzoin fumes. At any rate, this son was not permitted to return.

"For seven years the old lady's body lay there. Neither

side would yield. Finally we Dutchmen had to make an end of it. The pretender was permitted to come to Karang Asem for his mother's cremation. A great cremation it was, too.

"But it was our talk about hand-to-hand fighting that made me think of that. Something happened that day; I never knew what it was all about. There was the man from West Bali, with his lords all about him, magnificent with golden *kains* trailing from their armpits, and at the back of each was the gleaming hilt of a *kris*. Suddenly, in all the excitement and swirling of the crowd, something startled them. In an instant they were back to back, in a compact circle, and in his right hand each man had a *kris*.

"It was all over in a minute. Everything went on calmly as before. I never knew what had aroused their fear. But never will I forget that sudden, compact circle, bristling with slender steel. They were fighting men."

"What changes time brings!" I observed. "I feel it especially here at Tampaksiring, with all the ruins that are hereabouts. The mystery is a vastly fascinating thing.

"Those tall cenotaphs, especially, against the cliffs in the canyon of Gunung Kawi, and their rock monastery. What great empire may have centred here?"

"Only one of the ten has an inscription," said the Dutchman. "It says 'Here lies King——' but the name is gone. And the experts can't agree whether the writing is of the Fifth or Eleventh Century."

"And that great bronze Chinese drum in the temple down the road. It must be eight feet long and six feet in diameter. So heavy a drum was not a war drum surely; it must have been a gift."

"The natives say it is an ear-ring of the god who made Gunung Kawi," said the Dutchman.

"A Frenchman who had lived many years in China was with me when I saw the drum the other day," I went on. "He said the characters on it were Mongolian. But it was here when Dr. Rumphius visited the island, three hundred years ago. Even then it was ancient. Could it have been before the Mings? Could it have been a gift of Kublai Khan to a Raja of Bali?"

"Oh, yes, changes come," said the Dutchman, "and perhaps it is for the better. Now, for instance, these people are so priest-ridden. They make themselves so poor spending money for the temples, for cremations, for feast offerings. That cremation at Karang Asem cost the Government 25,000 guilders. Ridiculous. Really the people would be better off if they did not spend so much money for these things. It's been made the law in North Bali now that they can't spend more than 4,000 guilders for a cremation, or 40 *ringits* for a wife.

"It will be better when they've learned not to spend all their money for such things. It will be better when they've learned to buy more clothes. Good for trade also.

"For that matter the natives in Java have a pretty good lot, now that their island has been developed. They are not independent like these in Bali, but they are well taken care of. They have good clean homes in the plantation villages, and doctors to see that their health is good, and—"

I did not hear the rest of what he said. There was a lull in the chatter at the table. From deep in that chasm before the rest house, down by the temple, I heard the dim reverberation of a gong.

XXVIII

WHEN war came to Europe, a blond, blue-eyed German youth, just turned military age, was studying music in Russian Poland. Frontiers became barriers overnight. And while his kinsmen marched to death upon the Marne, Walter Spies * went into the stillness of internment in Siberia. Strange child of fortune, he dwelt five years in a lonely town beyond the Urals, the pampered prisoner of the Czar and the Revolution—fed, clothed, left to his own devices beyond the tumult of the world's frenzy.

Sometimes he tells of those days. Springtime on the steppes, broad miles of cobalt flowers; a week later, and the plains all garbed in yellow. Butterflies. Days with pallet and easel. Days with the piano. Rocks upon the heights, with rich luminous crystals and sea-born fossils. Flat plains, that changed their levels in sudden precipices; winter snows that levelled everything; and houses beside the cliffs, snowed under, breathing through icy tunnels. Tongues to be learned, Russian, Tartar, Persian. Books. Long winter nights with the lore of lands beyond the

* Pronounced *Speece*.

Himalayas. Such were the war and the Red Terror to
Walter Spies.

Peace, and years of wandering. Five years in Java.
Communion with the cloistered art of Jokjakarta, study
of the strange, exquisite music of the *gamelan*. Concerts
with five pianos strangely tuned. Dreams of Majapahit.

Spies came to Bali, stopped for a week-end, looked,
and listened. The next week he packed up his goods and
moved to Ubud, in the South. For a hundred dollars he
built a bamboo house. He took there his Persian rugs,
his easel, and his piano. He would dwell with these people,
learn their secrets of life. He would paint them, and
sell a picture now and then to pay for rice and macaroni.
He would record their rich lore of design. But above all,
he would put in black and white their music, whose elab-
orations of tone had been carried only in their memories.

One day, and it will not be many years from now, all
that will be left of Bali will be in Spies's records. Otherwise, all the best of it will be gone.

Mr. Hoover is abolishing poverty, isn't he? The De-
partment of Commerce must find wider markets, mustn't
it? Surely Bali must be made shirt-conscious.

The Resident at Singaraja and the Governor General
at Weltevreden are enlightened men, and they hold the
predatory off from Bali. But today's despoilers come as
Greeks bearing new-taught wants—to trade for birth-
rights.

Spies is writing down Bali's birthright.

There is Spies, standing beside his gate, a tall gangling
figure with khaki pants a bit too short and thick yellow

hair fringing below a flat felt hat. Deep eyes dominate his gaunt, dreamer's face. He is smoking a cigarette.

"Will you smoke one of these?" he asks.

It is a pinch of hairy tobacco, tucked in a slender conical roll of corn husk, held with a bit of thread. The corn husk is sweet. The smoke which trickles through the cone smells like the very devil. But it tastes pretty good.

I am sitting with Spies, listening to music. Concentrating, trying to understand it.

"I can understand," I say after a while, "why some people consider this music monotonous."

"Monotonous? Is the design of wall-paper monotonous? Is that changing rhythm monotonous, or the way that theme is passed about?"

We sit listening. Tone spangles the tremulous air. Spies sits in pensive, rapt attention.

"How long did you intend to stay in Bali when you came here, Spies?"

"Twenty years . . . Listen . . ."

He purses his lips and whistles. From the bewildering web of sound he plucks a delicate thread of melody.

"I never heard that theme before . . . It is beautiful . . ."

I encountered Spies at the feast of Tampaksiring.

"We had shadow plays at Ubud last night," said Spies. "They were beautiful. They are having them again to-night. They are celebrating somebody's birthday; I don't know whose."

So we drove down to Ubud, after a day of festal glamour at Tampaksiring.

There is no twilight in the tropics. The sun rises at six, and at six it swoops down past the horizon. You can see it moving, and suddenly night has come. But as the shadows lengthen there is a witching hour when the leaves yawn and relax in relief from strenuous heat, and all the world exhales a faint perfume, indefinably blended from a hundred kinds of foliage. From the rice fields of Bali troops of ducks romp homeward, before the bamboo wands of herdsmen.

Dragon-flies drifted overhead as we rolled down from Tampaksiring. Fey female figures poised on the paths that edged the water terraces. They carried long slim wands, waving them overhead, swinging them low over the rice leaves, as if to weave enchantment. What were they conjuring? Their wands were forked and wet with glue, and gossamer wings caught on them. They would take the dragon-flies and fry them in coconut oil. They were very good to eat, said Spies. I wanted to taste this sprites' ambrosia.

Now suddenly the women vanished. The sky was cavernous and glittering. The watery hills were lambent with starlight, as we rolled down to Ubud.

Spies sat at his piano. His long, slender hands were overlapped on treble keys, flickering. It sounded like harpsichord music, or an antique suite.

"This is part of that piece you heard the Belaloan *gong*

play the other night," he said. "It would take five hands to play the whole score."

"I had two members of the Belaloan *gong* staying here for a time," he went on. "I played them some of our occidental music. I don't remember exactly what the selections were. But I played one of Ted Lewis's jazz phonograph records. They laughed delightedly and thought it was fun. Then on my piano I played a bit of a Beethoven sonata. They said there were too many melodies twined together and one after another. They could not understand it. I played them a Chopin prelude and one of those little things of Mendelssohn's; they said both sounded like Malay opera—you know, those sentimental, yowling, westernized tunes that you hear on the Chinamen's phonographs up at Buleleng.

"Then I played them a Bach prelude. That, they said, sounded like music. That was something they could understand."

I heard this with a delighted shock, for it proved to me that I was catching at least some of the spirit of this exotic music. A few days previously I had written in my notes: "It is my belief, which I have not been able to test, that the Balinese musician would listen unmoved to a prelude of Chopin, find naught but nonsense in the Mendelssohn concerto, but find affinity in a fugue of Bach."

Spies's fingers drifted into something else, a blithe passage in a major key.

"What is that, something French?" I asked idly, for no particular reason.

"It is a bit of shadow play music," he said. "They have a different *gamelan*."

The shadow play would not begin until late. In Java it would start at sunset and last until sunrise, with one man speaking, working puppets, for the whole twelve hours. Here, it lasted only three or four hours. The shadow plays were better in Java. They had come down from past ages, with their strangely grotesque figures, with long noses and gaunt jointed arms that dangled below their knees, telling Hindu myths of the Ramayana and the Mahabarata.

Meanwhile we strolled down the street. It seemed thronged with people, blending in the darkness. From behind a wall came a man's voice singing, a weird wailing tune. We came into an open, flat place, the cremation ground. Oil lamps were hanging there, and there were scattered groups of people. There was a little house on high stilts, ornamented with coloured paper, and a steep ladder leading to a little door.

"They are going to have the Witch's Dance," said Spies. "Months they have been preparing it, but every time they planned to have it, it rained. We can come here before going to the shadow play. But there will be a long wait. We had better eat."

So we went to dinner.

"You had better drink plenty of coffee," said Spies. "You have a hard night ahead."

XXIX

ONE morning last September in a certain town in South Bali (recounted Spies as we drank our coffee) a young man came to his friends at sunrise with a strange story.

His eyes were bloodshot and he trembled as he spoke, for his night had been sleepless and he was very much afraid.

During the night he had awakened with a weird, uneasy feeling; and, unable to sleep, he had gone for a walk. As he approached the cremation ground he saw a light. Coming nearer, he found it to be the headlight of a bicycle, which was running around in spirals on the field of death. It had no rider.

Drawn irresistibly, in an agony of terror, he went still nearer. Then he saw that the bicycle was breathing. Its front tire swelled and contracted in regular respiration. With the courage of panic, he rushed upon it, and drove his knife into the pulsating tire. There was a terrific explosion, and a burst of light. He ran homeward and never once looked behind him.

So the men of the village went to the cremation ground.

There they found the dead body of a woman. Her breast was laid open with a great knife wound. They took her and buried her, and said no word to the Dutch authorities. The young man who had slain her became a hero in his town.

For this woman was known to be a *léak*. She was one of those obscene creatures who turn themselves into strange shapes and haunt the night. Now her iniquity had been proved, and no longer might she make the dark hours frightful.

Hard to believe that? Listen then to the tale of the nephew of Jokorda Ngura, who is a high caste man of Ubud.

An heirloom in this young man's family was an ancient palm leaf book, which told how to become a *léak;* and his father gave it to him. The son undertook to study its mysteries. He went to the temple at night, when everyone else was sleeping. There he drew a circle on the ground, and made marks within it, and uttered certain incomprehensible incantations.

Suddenly he found himself standing knee-deep in water. He walked homeward, and everywhere he stepped in water, until he entered his father's gate. Then all at once he stood on dry land, and his *kain* and feet were dry.

Other nights he called upon the occult powers of his book, and many strange sensations came to him. Sometimes as he went home he felt that he was flying, or crawling, or swimming; and there were other mysterious feelings that he could not explain. But always he could see his own hands and feet and body, and recognize that no change had taken place in them.

One night, however, all his incantations seemed to no avail, and at last he gave up and started homeward. As he left the temple and started down the road, he met two men whom he recognized. They stopped, stared at him an instant, and then with shrieks of horror ran away. Somewhat puzzled, he continued to his home, and slept well.

In the morning, he met the Pungawa, whom you probably will see tonight.

Had he heard, asked the Pungawa, of the frightful experience which befell two men of Ubud last night?

The young wizard started, for they were the same two men he had met. But no, he had not heard.

"They were walking near the temple, when the moon was high," said the Pungawa, "and they met a *léak,* in the shape of a great, monstrous, black dog."

Whereupon the Jokorda's nephew lost all enthusiasm for the black art, and hid his book away, and did not visit the temple any more at night.

I have seen a *leak* (Spies continued).

It was at the burial ground of Peliaton, near here, where bodies are put to await cremation. A woman had died just before her child was born. She was not buried immediately, for the people knew that in such cases the child may be born after the death of the mother. So the body was laid out on the burial ground, underneath a shroud. This, then, was the time for *léaks,* for they flourish in breathing the stench of death.

Nearly a fortnight after the woman's death I was sit-

ting here in this room with Dr. Rulof Gores. He is from
Java, an archaeologist in Government service. With us
were the Pungawa and two sons of the Raja of Ubud.*

A boy came in very much excited. He said a *léak* had
been seen at the burial ground of Peliaton. So we went in
Dr. Gores's car to see.

We left the car two hundred yards away and walked to
the cemetery. It was deserted, but the night breeze told
that the body was still there. About forty yards away we
sat down to watch.

It was a moonless night, but the brilliant stars cast a
dim sheen over the whole world. Beyond the palm trees
which encircled us, miles of watery *sawahs* cast back reflections of the heaven's glow. We sat there silently for
half an hour and nothing happened. We got tired of waiting. We were just deciding to leave when I saw something.

Then we all saw it, out among the *sawahs*. An eerie,
cold violet light, which moved up and down and in circles,
and cut strange patterns in the darkness. It pulsated, as if
breathing, and slowly it came toward us. It came up over
the rice fields, moving in weird curlicues, until at last it
poised over the woman's corpse. There it hovered, pulsating all the while. It expanded into a great, dim, luminous
mass. It contracted into a tiny concentrated gleam that
dazzled my eyes. Yet there seemed to be no reflected light
in its vicinity. Above and below it, and on all sides, I
could see the distant landscape. Nothing was to be seen
there, except that ball of light.

After a while we decided to go closer and investigate.

* Precisely the same story was told to Roosevelt by a son of the Raja.

As we approached, the light grew dim, and when we came close it vanished entirely. But when we went away, it appeared again. Three times we did this, with the same results.

Again we watched from a distance. It hovered a while longer over the body, and then moved away, slowly as it had come, and disappeared among the palm trees.

We were pretty nervous by this time. We went back to the car in a hurry.

But yet another shock awaited us. For in our absence the car had been piled with rubbish. The seats and floors were covered with old rotten meat, putrid sausage, banana skins, grass, and clods of earth.

That doesn't seem to you the sort of thing a supernatural being would do, does it? Well, neither is it the sort of thing a man of Bali would do, to a car being used by the Pungawa and sons of the Raja.

We went home, but even then the *léak* of Peliaton was not done.

For in the morning a good wife of the village went to a spring near the burial ground to fill her water jar. As she leaned over the water, she saw a woman sitting on the bank, laughing at her. The woman rocked and shrieked with laughter, and not a sound escaped her. Her eyes were wild.

The wife was afraid and ran to tell her husband. He came and found the strange creature still there. He cried to her, and she answered with silent guffaws. She moved nimbly through the palm trees to the graveyard, and sat upon the shroud. The man ran to the *balé banjar*, climbed to the bell tower, and beat a mad assembly call on the hollow hanging logs.

Two hundred villagers came rushing to the burial ground, and still the woman sat upon the shroud, laughing, laughing, laughing, a young woman, and her long hair hung in wild profusion about her shoulders. The throng circled around her and strained its ears for the silent peals of her ghastly mirth. They shouted, and their voices echoed. Her beautiful brown breasts shook in obscene merriment.

Numbers make strength and courage, so a few bold spirits stepped forward from the circle and advanced upon the mad creature. As they came she moved aside and lifted the corner of the shroud. They came closer, and with a sudden motion, she rolled herself beneath the cloth. Now the crowd surged forward. The shroud was seized and pulled aside.

Nothing was there save a human body, two weeks dead.

They buried the dead woman that day, with her unborn child.

XXX

THE Witch's house stood on its stilts at the end of a space ten by twenty yards, bounded with upright poles. The poles were festooned with banana leaves, since this dance usually is given in a wooded place. Cords were strung between them, from which hung flickering oil lamps. It seemed as though all the village must be grouped about this area—men on one side, women on the other, children everywhere. But as many more were down at the other end of town, waiting for the shadow play.

We sat down with the men and waited. Behind us were the ever present dim lamps of the little market girls, with their stands of peanuts, siri, coconut drink. Beside one stood a little naked boy, eating his supper of rice from a banana leaf. Beside the Witch's house a *gamelan* was playing.

This was its theme:

It was the same theme over and over again, and about it entwined elaborate ever-changing embroideries. I sat there learning how to listen to it. These people seemed to pay it no attention at all. It was there as a background of design for a glamorous occasion, like the backdrop of a revue. It was perfect for the purpose. If played over the radio, it would be the ideal defense against those misguided hosts who turn on the loud speaker during a bridge game.

We sat waiting for a half hour. One learns to be patient in the tropics. Spies was talking with some natives beside him. Suddenly he exclaimed in chuckling delight:

"They've changed their minds. They're going to have the *barong!* The *barong!*"

The theme of the *gamelan* changed, and its rhythm was a *scherzo*.

A fierce apparition burst into the arena. A great, glaring mask, with fierce teeth and a flowing lion's mane. Its jaw was hinged, and snapped with savage impact. It had the body of a beast, with long hair and golden trappings, spangled with little mirrors. It had a long, thin tail. Its paws were two pair of human feet. The *barong!* I had seen it ambling along a road. I had seen its trappings hanging in a temple. It was very sacred.

It trotted about the stage, lolling its head from side to side, glaring, snapping ferociously. Its watchers rocked with laughter. Sacred? Surely. But what man loves a doleful god?

The beast walked with mincing steps, and now with stately strides. It swayed, hopped and glided from side to side. It craned its neck, and peered down sidewise, inquisitively. It collapsed like an accordion, it stretched to its full extent, lifting its head majestically, and wiggled its hind quarters. Its tail wagged vehemently. Did you see Charlie Chaplin's biscuit and fork dance in *The Gold Rush?* Here was the same spirit, gone oriental. The sidelines rocked in ecstatic gusts of laughter.

Now pathos. It was a wan, dejected beast that plodded here. Its tail dragged. Its jaw dropped and softly closed. It slowly turned its face toward us. Was it possible that this fierce, rigid mask was looking wistful? What racial genius had contrived this strange, dramatic pantomime?

Anger. Rage. The lowering mask turned full upon the *gamelan.* The creature reared, crouched, intent upon a newfound foe. It catpawed forward. It pounced, charged, swerved, retreated, charged again. The music went on as before. The beast tossed its mane, ground its teeth. And now, amid a burst of hilarity, it turned its tail and vanished through the crowd.

"They decided not to finish the *barong* dance," said Spies. "It might have gone on all evening. But they want to have the Witch's Dance. It may rain tomorrow night."

At the far end of the arena appeared a tiny dancer. It was eleven o'clock.

The dancer was a boy nine years old. His head dress

was like a *legong* dancer's crown. He wore a shirt with tight white sleeves. His *kain* was blue, and about his waist and chest were wrapped several turns of light green scarf. He danced with his hands and eyes, as in the *legong*, but his feet moved in long smooth glides, weaving from side to side of the dancing space, on heels firmly planted. Now he moved swiftly, now froze in a weird pose of expectation. Now he spoke, or rather, sang, a short, plaintive measure. The natives burst out laughing.

Without a sign of recognition, the boy continued dancing. Again he sang, the same brief, minor tune. Again the laughter. It was not hilarity, yet it did not seem cruel. But something was amiss.

"Why are they laughing?" I asked. "He is not a clown."

"He held that last note too long," said Spies. "In this dance are the same tunes as in the *arja,* the dance drama, and the people know them. Each dancer has a different tune, always the same. Each time the dancer speaks, it is the same tune."

The boy danced on for twenty minutes. Then appeared two other dancers dressed like him. They danced in the same manner also, while he rested before the Witch's House. They were girls. They sang also, each with her own peculiar tune. Now I knew why the crowd had laughed. There was a cameo precision about these little songs. Like the dancing, it was a set, stringently disciplined art form. Not that I think Westerners would like this music (though many occidentals have learned to like the taste of Bombay duck), but oh! how it contrasted with the "cuteness" of our own juvenile performers.

The dancers came and came. They were the Witch's pupils. Evil spirits, flickering in the lamplight. Singly and in pairs they postured, and each seemed better than the one before. They were more skilful than Runis and Madé Réi. It would have been monotonous to watch them all the time. The natives did not. They sat, and watched a bit, and chatted among themselves, and the dance went on and on. In the arms of the man beside me, a little boy had nodded off to sleep. Down by the Witch's ladder, the First Pupil, butt of laughter, had laid his golden crown upon a friendly adult knee, was dozing. The seventh dancer finished, and rested with her fellows.

Tension. Expectancy. The baby beside me was awake and staring. The dozing dancer sat up, rubbed his eyes.

Shade of Endor! Vision of the Brothers Grimm! There was a witch! Witch of the East, of the West, a witch the world over.

Her face was chalk, with evil lines of black, and down beside it straggled long grey hair. From head and shoulders a long grey shawl, a brown *kain* trailing on the ground. A figure gnarled and crouching, right hand weighing on a crooked staff. Left hand upraised, weaving patterns with long, sharpened, obscene nails.

She hobbled, stumbled, swayed; she sang an evil tune. Her voice belched out in gutturals. Insinuatingly she made her way, eyes glaring in malevolence. Pupils knelt before her. Double double, toil and trouble. Words with rasping intakes of the breath. Power of darkness, sage of evil, teaching lessons.

On she went, and up the ladder to the little house, and closed the door.

All the while the *gamelan* playing, and in the crowd a man's high nasal voice, singing an incantation, like an oboe.

It was one o'clock.

"Would you like to see the shadow play now?" asked Spies.

No, shadow plays could wait until another time. I would see this through.

Comic relief. Clowns with painted faces, scraggly beards. Mincing gaits, waddlings, straight from a burlesque wheel. Low comedy, the same the world around.

Clown in a woman's wig, stomach padded, rolling on the ground in agony. Clowns running about in desperation. Quick! quick! call the doctor! Savant of herbs and potions, mystic passes with the hands. Buffoon in mock birth-pangs, audience hysterical. Low comedy, the world over, out of the aches of life.

Memory from a theatre in Singapore—poor woodcutter, dead broke, and his wife in burlesque contortions, then running to the wings to fetch a wooden doll. Night of Japanese melodrama in Kyoto: heroine and audience in convulsions and my fat little rickshaw boy beside me (Ké, his name was) slapping my back with a shout of glee: "Now! she is making children!"

Meanwhile a man was digging a hole in the dirt before the Witch's house. He retired and came back with a young papaya tree, freshly cut. He planted it, and with his feet tramped down the dirt. That made the place seem more like a forest.

Now entered a tall male figure in a golden crown, son of a Raja, come to exorcize the evil ones. Chin high, knees bent and pointed outward, he advanced with stately tread. His legs were encased in tight trousers, and from his arm-pits, like a cape, hung a crimson *kain,* stiff with gold leaf. Daintily, as he danced, he lifted his cape, like a lady pacing a minuet. Something of a popinjay, conventionalized representation of royalty. At his back a jeweled *kris* glittered.

He spoke interminably, in a querulous falsetto, of irritating affectation. The natives could understand no more than I of all this preciosity, for he spoke an ancient, forgotten tongue of aristocracy. Now his two retainers took up the talk in common speech, gruff and uncouth in comparison, making known what he had said. Nothing so much distinguishes the brown man from us of the West as his lack of restlessness. We demand new words, new tunes, new action all the time. He will sit for hours in perfect patience.

But now there was action. The Prince was challenging, before the Witch's house. A Pupil came to meet him. She darted, circled, she scolded him. He advanced in haughty menace. She lunged, arm straight in the attitude of killing. He retreated, his slender *kris* was shining. Around the papaya tree they wove. Again the Prince came forward, and seized the hand she stretched out toward him. He whirled, and spinning in a circle the tiny sprite's feet spurned the ground, faster, faster, furiously. And when his elfin foe was vanquished, the Prince stood scornfully, waiting.

The door of the little house was opening. A burst of smoke, a shower of sparks and coloured paper; the pop and fizzle of firecrackers split the silence. There amid the fumes now stood the Witch, incarnated as the torch-eyed Rangda. Rangda the Widow; Rangda, the evil one. Her long red tongue rolled thirstily between her sabre teeth.

Rangda descended, and all about her great round head were hanging shreds of furry hair. The Rangda danced, a lumbering uncouth dance. Her red visage bent toward the earth, it turned upward toward the heaven, and above were lifted savage, long-nailed hands. A deep roar rumbled from somewhere, and bladed hands were converging toward an awe-struck Prince. He crumpled down before them.

A dead Prince walked away unheeded, and the Rangda danced triumphant.

Until now all dancing had been of a certain studied pattern, schooled to a fastidious refinement. But here was something primeval, that burst beyond all bonds. It glared from antiquity, like that white figure of Tintiya that had gleamed from a shadow in Spies's house—Tintiya, a tribal god, become omnipotent. Rangda was dancing, a rolling, bumping gait, feet lifted high. And all the while were deep-voiced, moaning growls.

The Rangda vanished, down through the crowd, toward the temple.

A bare-chested young man, with aristocratic profile, stood beside us. Spies introduced him, the Pungawa, Chief Magistrate of his district.

"The shadow play was excellent," he said. A cultured gentleman, a connoisseur. "I saw the latter part of this. It was not so good. The actors of the *arja* are much better."

The shadow play! I had forgotten it. And now it was three o'clock. The night was over. But no, it was not over. The crowd, men, women, children, was undiminished, and down there by the temple something was happening. We went to see.

There was the Rangda, growling, moaning. There was the *pumanku,* keeper of the shrines, in his white *kain.* There were holy water, incense, palm leaf offerings. The Rangda lifted up her ghastly voice:

"Come hither! Come hither, all you evil spirits!"

And now the Rangda, reeling back to the dancing square:

"Where are you? Where are you, all you ghosts?"

The Rangda moaning, fangs and fingernails, and all about her seven evil spirits, flickering white and gold.

The spirits filed away. The Rangda stood, and men rushed to her. They lifted the heavy mask.

There stood a young man, with eyes entranced and glaring. He swayed against supporting shoulders, and rough air rasped between his open teeth. Was that a fleck of foam upon his lip?

The Witch's dance was over.

The moon had risen, and as we strolled homeward the night was Gothic. Ethereal gargoyles filled the shadows, and this palm-lined road was a pillared nave, vaulted with leafy silver.

Rest, from a day of dreams and visions; sleep's oblivious shroud descending, shutting out a wakeful world of phantoms, nightmares. Drowsing, I lay on a friendly couch, sinking into soft forgetfulness.

"Listen," said Spies. "They still are down there. It's half past three o'clock, and they still are playing."

From far, far away came the muted ringing of the *gamelan,* but now the voice of Spies also seemed very, very far away.

XXXI

WHAT a day it was, my last in Bengkel. It was the birthday of Ida Bagus Putu Yappa, Gidéh's little son. But it was more than that. Sayu, she of the captivating regal air, had conquered the Padanda's house, and now was welcome there. It would not have been tactful, pointedly to hold a tardy marriage feast. But over this day's festivity hung a joyful air of reconciliation.

The *janger* had prayers that afternoon, under the trees in the little temple at the Padanda's house. Beneath his shelter Ratkarta was muttering and making mystic signs with his hands, and his little brass bell was ringing sweetly. All about the temple the thirty-five youths and maidens of the dancing club were chatting, laughing, smoking cigarettes as they put on their costumes.

Renang and the other girls bound their breasts in green and yellow scarves, and around their waists put girdles of bright tin. Through their ears were tubes of silver, studded with gold; and their brown hands and faces were dusted with a tinge of saffron. Kumis was painting their eyebrows with a black grease pencil, shaping them with a razor blade. Their foreheads were painted in the sem-

blance of black parted hair, and upon this paint the regal
Sayu was sticking fine cut patterns of white palm leaf.
Their heads were crowned with shivering sprays of white
and yellow flowers, imbedded in a bit of coconut husk.
Their hair was heaped high behind, and its trailing ends
were hung with frangipani blossoms. What an airy sprite
was Renang that day, as she sat amidst her blithe and
dainty group.

When all was ready they knelt before the shrine, and
the old *pumanku,* assistant to the priest, came with holy
water. He sprinkled them all with it. He sprinkled all
their clothes and the properties of the dance, the horrid
masks of the Rangda and the Garuda Bird, the false moustaches of the boys and their Moslem fezzes. He poured
holy water in their hands and they drank it. He put flower
petals in their hands. They placed the petals on their
brows. They touched their hands finger to finger and
bowed their heads in prayer.

And now before me a miracle was happening, an instant's essence of the fathomless East. The *pumanku* was
burning incense, and into its slowly curling fumes he
stared. A silence settled over everything. The Padanda's
bell was still. The *pumanku* crouched there behind his incense bowl, motionless as an image. His eyes glared and
his lips hung open. Now suddenly he was speaking. His
voice burst forth in loud gusts, as if beyond control. He
sat transfixed, and not a muscle moved. His voice rasped
and jolted in the stillness. Mystery hung thick about us.

"What is he saying?" I whispered.

"Sh-sh," said Réné. "The god speaks."

The *pumanku* squatted there, graven as if in stone. On ran his voice and a hundred pious ears hung on it. I breathed stealthily, stood tense, burst into a torrent of sweat. At last the rumbling, entranced voice was still. The *pumanku* closed his eyes, and slowly rose to his feet.

Now everywhere were smiles and laughter, and tense excited talking. Everyone was speaking, grinning happily. Roosevelt was chattering with Réné.

"What was he saying?" I demanded.

"They tell me the god was speaking," said Roosevelt. "The *janger* is looked upon with favour. If they practice much, they can be the best *janger* in Bali. But they must work hard."

The youths and maidens danced that night, in a little open space outside the Padanda's house, to celebrate his grandson's birthday and the peace of his household, and to justify the favour of the gods. They danced sitting under a square canopy hung with coloured paper and dim lanterns, and close grouped about them were all the people of the village. There had been rice cakes to eat in the temple, Ida Bagus Gidéh had taken us aside in his house, and had given us fine pieces of juicy duck. His brother Ida Bagus Rei had come with a great bamboo of *tuak,* and we had drunk our fill. All the village was mellow with this sweet palm wine, and now was grouped around the dancing square.

See the *janger* dance. Shoulders twitch with the rattle of the drums, and the flowers of each head-dress shiver.

Girls' heads slide sidewise on their dainty necks, and their black eyes snap. They sing a lilting minor tune, and their white teeth flash. They dance with their heads, with eyes and shoulders, with swaying torsos; their arms and hands swing like wavelets in a sudden squall, then like the billows in a calming sea.

Now, from the centre of the swaying line, one dancer slowly identifies herself. Gradually emerging from the sitting chorus, she rises to her knees, now to her feet. Renang is dancing; Renang, the best of all! She crouches, knees flexed and shoulders forward, and the fan in her extended hand vibrates in a blur. She advances on heels with toes well out, and her feet beneath the trailing sarong are sixteenth notes in a rapid beat.

Renang is fourteen, soon ripe for marriage. I saw her weaving in Kumis's house this morning. She was retired last year from the dancing of the *legong,* too old, surely, for its flickering speed. But Renang will show them. Can this be she? Her waist is quicksilver and her feet are swallow's wings. Freed from the classic restrained measures of the *legong,* she dances the dance of Renang, in which is all the dance of Bali. She gilds this popular *janger* with a tinge of consummate art.

She crouches, in rhythmic postures beyond all probability, she stands erect in diminutive grandeur. Her hands are like the fairy branches of the frangipani, poetry of angles, mutable geometry. Each motion is part of a balanced surprise, implausible Q. E. D. of equilibrium. Her dance is of human lines compacted in a sonnet, as sexless as a page of corollaries.

Now her steps are longer. She weaves amidst uneven shafts of light. She is a faint dimness in the shadow. She is a shining, glaring goddess. Her sharp-cut little face is graven ecstasy in stone.

What can it be, this fleshless motion? What is the eerie spell it casts? Whence creeps this strange paralysis upon me? I watch Renang dancing and I scarcely see her. My lungs are stifled in the breathless night.

Now before her vacant place she falters. Slowly she melts to swaying anonymity. Now there is a row of those tireless, singing, smiling girls and shivering flowers. Ida Bagus Gidéh and Sayu standing there, can you tell me which one of those is Renang?

XXXII

THE *janger* danced that night as it had never danced before, and now Sayu retired with her colleagues in wifehood, and Ida Bagus Gideh with his younger brother Rei came to our *balé* to call on us. They brought a big bamboo of *tuak,* plugged at the mouth with coconut fibre to strain the dregs. It was cheaper that way; by the bottle it cost all of a cent a pint. It was good, this sweet cider of the sugar palm, that was eighteen hours old. So we sat sociably about the table in our *balé,* and Kumis was the keeper of the *tuak.* And all about us was a ring of friendly, handsome faces, as we talked of many things, and Ida Bagus Gidéh told us a story.

"Many, many years ago, long before the white man came," said Ida Bagus Gidéh, "our whole world, as far as men knew of, was under the realm of Majapahit. In all the world the Raja of Majapahit was the greatest. His city was a great city, with high walls and watch towers, and all the routes of the world led to it in Java. And the Raja Browijaya was the greatest that Majapahit had ever known.

"But one day his chief Brahmana came to him in fear and trembling, bearing a *lonta* of sacred script in palm leaves. For in this book he had found a prophecy, which told that in forty days the name of Raja of Majapahit should be finished and there would be no Raja any more.

"And when Browijaya heard the prophecy, he feared it. The thousands of men in his armies he set to bringing wood, until they had built of it a vast pile. And when the fortieth day had come, he mounted the pile, and there at his own order was burnt alive.

"Now Browijaya had a son, who was named Dewa Agung Ktot, who also heard the prophecy of the book. Ktot gathered his Gustis about him, and their men, and sailed from Java. At Klung Kung he established himself, supreme among the seven Rajas of Bali."

We were all silent for a moment, but then I spoke. Why, I asked, had Browijaya believed the book? But as Ida Bagus Gidéh pondered his answer, his brother broke in upon our talk. Ida Bagus Rei was feeling merry, and had small patience with this solemnity. Had he not brought *tuak* for us?

This was my first meeting with this gay young Rei. He had been away to college in Java, and wore a striped shirt in token of it. He was very amiable this night.

Was it true, he wanted to know, that I was going away tomorrow? When was I coming back? I must come back soon. When I came back he would have a fine roast pig for me, stuffed with coconut and peppers. For me he would have a whisky and soda; but for himself, he would have *tuak*.

So we had another glass of *tuak* all around.

Rei was the first Balinese I had met who knew any English. He had picked up a few words from an Armenian trader. And now, to my surprised delight, I found that I was really conversing with him. A few words of English, a few words of Malay, and a few drinks of *tuak* make a fine conversational medium. I found myself telling him about America.

In my island, which was New York, I told him, Tuan Roosevelt was an *anak agung,* and a great *tuan.* In fact, in my island, there was a Hotel Roosevelt, a *passangrahan,* which was a hundred metres long, and seventy-five metres wide, and more than seventy-five metres high.

Rei gasped. He exclaimed in admiration. He called upon all the others to witness the greatness of the *tuan* that was among them. He poured another drink of *tuak.*

I felt rather flushed with success. I thought of other things to tell him. The subways that roar through the earth with millions of the mashed. City Hall Park at sundown, and over it the Woolworth Tower, and Nassau Street belching its tired thousands. The streets of my island, where I might walk for miles and never see a blade of grass, nor smell a petal. The population of my tiny island, many times that of all Bali. The marvels wrought for human comfort by our engineers, who had made it possible for hundreds of people to live in layers, in less space than was taken up by this, Kumis's house, and to live so away from all the smells and sights and sounds and inconveniences of the country-side, right close to the machine-work of their waking hours. The theatres of my island, where men paid five *ringits* to see girls glorified. But I decided not to tell him these things. After all, my vocabulary was limited. And I valued Ida Bagus Rei's friendship; I did not want him to think I was a liar.

As a matter of fact, I was tired out, and drowsy with drinking. I had got up at dawn. It had been a hard day. Now it was very late. Though the Balinese never know when to go home and seemingly never to go to bed, I fortunately had learned something of the excellent manners of Bali. Since I wanted to go to bed, I went.

As I bade goodnight, Ida Bagus Gidéh spoke to me.

"You wondered, *tuan,* why Browijaya believed the prophecy. You have been in Java. Did you see there Majapahit's city? Tomorrow at Klung Kung you will see a crumbling palace, but you must look well to find a Raja there. Have you no prophecies in your island?"

It was good to lie down on my hard, cool couch. New York seemed very far away. Science, engineering, skyscraper beehives never had been invented. It was good to know that friendly living things were all about, cows, pigs, chickens. I even felt kindly toward that cock roosting beside my head, though I knew that soon at daybreak I should curse him. It was good to be here among these children of Adam, still in their garden, who had not eaten from the tree of knowledge of ugliness and beauty. Who did not feel the rapture of the moonlight, or see its silver on the palms: for things were always so. I felt that they were close to the heart of life, and I was with them. I was glad I had not told them more about my island.

Drowsily, as I lay there, I could hear Roosevelt speaking. I could not understand him. I could see a ring of intent listening faces in the lamplight. I gathered that he was talking about the war. Then, in the strange primitive idiom we used in this pidgin Malay, a few phrases sprang at me.

"Many men have got. Great fire. Boom! One second. No man have got."

A gasp of credulous amazement. Faces with wide eyes staring at each other in the lamplight. A burst of excited talking. What wonders these great white *tuans* had wrought! What fear had they of prophecies?

XXXIII

I STOOD on a ship's deck off the roads of Buleleng, and the isle of Bali slowly moved away. A thin green fringe was by the water's edge, and high jutted the slim black cone of Tabanan. A mist was blotting out the land, hazing my vision of this isle of dreams. Could it be real? Had I been dreaming?

"Roosevelt wants to be lazy! Ha, ha, ha!"

A big fat Dutchman had answered my silence. He had a thick neck, this Dutchman, which bulged over the high collar of his uniform. A sturdy, efficient man, I knew. I didn't like his laugh.

"Roosevelt says he wants to be lazy! That's a good one."

So I went down between decks and got acquainted with the cargo. There were five hundred cattle and water buffalo down there, and two thousand pigs in cylindrical baskets.

There were five hundred parrots also, gay birds of green, crimson, white, and saffron, crammed in little wooden boxes. Three hundred had died on the way down from Ternate, and many more would perish ere the ship reached Singapore. And I thought: The colour of Bali will perish also, for this is a white man's world, and the white man bleaches. I have found this wealth of the Indies; let me carry it away and share it, before it fade.

And so I sailed up to Singapore. I had my clothes starched and my shoes shined, and wore a boiled shirt at the Raffles Hotel. I had my watch repaired. I rustled baggage, and found out how great a crop of mould a bagful of woollen clothes can grow in four months of rainy season. I dickered with a steamship office, and with money changers. Oh, it is filled with dull detail, this business of being a white man. I sailed still farther west, and whiled away the time with auction bridge. And so I came back from the world of visions to the world of Things. There were many more Things than there had been when I left. There were talking movies, and one-handed dinguses for the telephone. I told a fair young woman I had found beauty, and she replied that her family had a smart new car.

I returned bursting with strange and glamorous memories. I told of them. Questions were asked me. My mind was whirling in these days, but nothing works so much toward clarity as the queries and comments of other minds.

I was telling of the deep intricate culture of these village folk, how every day I learned some new thing that

cast a futile beam of light into the vast shadows of what
I did not know, whose immensity—

"Have they got any education?"

I pondered. I should be able to answer that, for I had
education. My mind went back to a freshman lecture room
and a revered savant speaking: "Now don't neglect to
take cultural courses. Why, I had a letter from a young
man just the other day. He had been visiting a millionaire
customer whose hobby was collecting paintings. And this
young graduate wrote me: 'Oh, if I only had studied something about art! I lost the order!'"

And thinking of these matters I realized that the men
of Bali had no education, that they were downright ignorant.

I was telling of the simplicity of Bali's life, of frugal
living, and wealth that went for music, art, and pageantry.

"Why, I should think with all that money," someone
said, "that they would have more comforts out of life."

I was telling of a temple feast, and of thousands in
their daily life turning with childlike faith to face the
mystery of being.

"Isn't it strange," remarked a listener, "that the less
people know, the more religious they are. Is all that fuss
getting them anywhere?"

I was telling a young woman about the art of Bali,
which was more a matter of spirit than of works, telling
of delicate craftsmanship and of its fruits that were always being made and perishing, an integral part of life,
as pulsating and transitory as flesh.

"But do you mean to say that they let these things go to ruin, that they have no museums or anything like that? It seems so wasteful! It doesn't seem as if they are *getting anywhere!*"

I was telling a young man about music, about Spies and his tryst with beauty.

"Do you mean to tell me that he's perfectly content to stay there and just exist, never *get anywhere?* Now, if he were going to come back, and give concerts, and get rich and famous—"

Getting somewhere. Getting anywhere. You may imagine how these sayings brought me down out of my clouds. As for myself, where was I getting? I had been chasing the horizon. For a year I had followed the setting sun. And where did that get me? Right back to the place I started from, to an overdrawn bank account.

And so I settled down. I got a one-handed dingus for my telephone, bought a radio, turned to my typewriter and became a solid go-getting citizen in the world of Things. Survival of the fittest, is it? All right, I'll survive. For all I know, I am contributing to the general Prosperity. I wrote a piece for the paper about a poor little girl who wanted some pretty things, and turned bandit to get them, and that was printed on an inside page. I wrote a story about Mrs. Vanderbilt getting a new private golf course, and that was printed on the front page. Oh, I am working very hard, and some day I may get my boss's job. Then perhaps I can get a smart new car. But I have not yet had time to calculate where that will get me.

Mulling over these matters I stood at dusk on Brooklyn Bridge, late in autumn, when the lights were bursting out from darkling spires. I used to go there often in the days before I became a go-getter, to gaze upon the dreams of men, towering above a Babel. I found myself there on this particular evening, and all at once I heard a voice beside me: "Well, I certainly am glad to find you. I've been hunting all over."

Turning, I saw standing there Ida Bagus Gidéh. Somehow, it did not seem strange for him to be there, with his bare brown shoulders. Apparently it was not strange to anyone, for none of the passing crowd delayed to stare at him.

"Well, Gidéh," I said, "this is a surprise. I'm surely glad to see you."

"My body is asleep," he said. "While I was wandering around, I thought I might as well come over here and have a look. You had told us such wonderful things about your island."

I don't remember whether we were speaking English or Malay. But this night, for the first time, we seemed to understand each other perfectly.

"Well," I said, "you came at just the right time. The Philharmonic Symphony Orchestra is playing this evening. You love music, I know, and you must hear it. You'll find it quite strange, but very beautiful. You know, when I went away we had two orchestras, for which Mr. Mackay and Mr. Flagler paid the deficits, but when I returned I found they had combined them into one. You will easily see what an improvement that is, and how much better music will result. As a matter of fact we'll soon get this matter of music down to a real business basis. With the radio and the movietone and whatnot, we'll be able to get along with just a few really first class orchestras. Just think how many of these fiddler fellows that will release to do real honest productive work and how much money—"

"Really," said Ida Bagus Gidéh, "I'd like to hear your music, but I must be getting home. You know, it's entirely beyond me to figure out what hour it is in Bali, but it must be time for me to wake up."

"But Gidéh," I said. "You must see our Metropolitan Museum of Art. Someone gave it another million dollars just the other day. I don't want to be boastful about my home town, but this is getting to be the art capital of the world. Why, I was reading in the paper just the other day, the head of some art school or other said that New York was coming to be the real centre of art, that with all the masterpieces we are importing this will soon be the true fount of creative inspiration. I tell you, before long the Louvre won't be able to hold a candle—"

"Ah, yes, your art," said Gidéh. "Your men build tall temples."

"Temples!" I exclaimed. "Now, Ida Bagus Gideh, as a young man who is going to be a priest, you must let me show you the Cathedral of St. John the Divine. It's Gothic, which shows how high-class we are getting to be. They had a big drive, and the money for it was given by prize-fighters and—"

"Ah, yes, your temples," said Gidéh, and I noticed that

he was gazing toward the glowing towers of Lower Manhattan.

"I have to confess," he continued, "that I would have found you a lot sooner if I hadn't been looking around a bit myself. I have been seeing your tall temples and the hosts of people in them.

"You see, I ran across a chap from Benares, who is travelling in pretty much the same condition as I. He had visited once in Bali, and you know how it is when you meet a fellow who is familiar with your home. He'd been here quite a while, studying conditions. I was a bit worried about his body when he told me that, for fear he might be buried by mistake; but he explained he is a holy man, and has trained his body to go into long periods of catalepsy. Rather convenient that, to be able to travel as you please while your body remains at home, adding to your reputation.

"At any rate, he had been in Bali for the feast of Besakih, and as an orthodox Brahman was very much interested in that annual reversion to our ancient tribal worship of the God of the Mountain. He said he had observed the same phenomenon here, but more pronounced, one day in every seven. He said there was a strange native name for this seventh-day feast, 'day when camels pass through the needle's eye,' or something like that."

I was very much disappointed in Ida Bagus Gideh. In the old days, when we hadn't been able to converse freely except through an interpreter, he had seemed a very sensible fellow. But if he wanted to swallow all this nonsense, and preferred as a guide some heathen swami, above me who really knew what I was talking about, I was not the one to correct him. I could see quite plainly that he had fallen in with one of those harpies who are always waiting to fill tourists full of bunk, and it would do no good for me to try to put him straight.

"Well, you pious old aristocrat," I said heartily, to show there was no hard feeling, "if you're going to study religion here, you'll find that the Brahmana don't cut so much figure as they do down your way. Our priests are a pretty sorry, seedy lot, and they don't have much power, though they are pretty noisy."

"Now, my understanding was quite the opposite," replied the young Brahmana. "From what I saw of the Brahmana in these temples, they were surrounded by rich possessions and many men did their bidding. My friend from India said that just a few Brahmana ruled by their secret vedas all the land; that in late years their most powerful dogma—put forward through a spokesman— had been that silence is golden; and that, like ours, their symbol is the sacred Bull.*

"Really I was fortunate to meet that chap. He showed me just what I wanted to see. He took me in one of these tall temples to a shrine with great metal doors, and men with guns were guarding it. I was surprised to find it was not empty like our shrines. It was filled with piles of golden discs, and each was inscribed 'In God We Trust.'

"My friend explained, of course, that these were not idols, but only symbols. Symbols of a god with many names. Some call him Things, and some Prosperity. And

* Gidéh's visit was prior to October, 1929.

his worship is a race of getting something, getting somewhere, and the devil takes the hindmost.

"He took me to temples also which were not so tall, and in them were the whirr of wheels and the clank of metal. Their altars lifted tongues of smoke to heaven.

"In the morning I saw the millions pouring to these temples, and in this evening the same millions of natives pouring home again. I marvelled. I said to my friend from India: 'You know these people. What is the reason for it all? How can religion so dominate a people's life?' To me it all seemed very strange and complicated, and yet it was all so simple.

" 'They believe,' said my friend from India.

"Well, I must be going, but I'm glad I came.

"I never saw my island till this day, *tuan.* We understand, you know, by contrasts only. Never before I knew that Bali's air was sweet, and gently scented with the breath of growing things. I never knew that days were dancing, that nights were limpid silence, tinged with music.

"We have not many Things, *tuan,* to break in earthquakes and erode in rains. But we have what we need.

"We are not getting anywhere, *tuan.* A thousand years we have been where we are. And we are happy there.

"Today is savoury, flavoured with a million yesterdays. We taste today, nor gulp it hurriedly to grasp the minted ashes of tomorrow.

"You think it strange, *tuan,* that I with mischief in my eye shall some day be a priest. Always to you it must seem strange. My father told me mysteries I may give to no man, save my son; and these are secrets no white man could ever know or understand, for in him there is lusting turbulence. But with them I and my people shall live in peace, hoarding the living treasure of today.

"There will be feasting tonight in Bengkel, *tuan*. The *legong* will dance, the *gong* will play. The moon will wet the palms with silver. There will be pig to eat, and *tuak* to drink—"

"And Renang will be dancing," I broke in.

"Oh, no! Renang will not be dancing. For I have married Renang, *tuan*. That's what the feast is for."

NOTES, DIGRESSIONS, AND
AFTERTHOUGHTS

CHAPTER 1

The incident of self-slaughter in South Bali is given substantially as told by Gregor Krause in the preface to his excellent book of photographs ("Bali: Volk, Land, Tænze, Feste, Tempel," new edition published by Georg Muller in Munich, 1926) made while he was a government doctor in Bali two decades ago.

H. S. Banner, in "Romantic Java," (London, 1927) recites that the high-born men killed themselves inside the stronghold, and the war-like women ventured forth alone for their suicide. In this connection he quotes Thomas Cavendish's diary of 1587, telling of a Javanese custom in the days close following the Hindu era:

"The custom of the country is, then whensoever the king doeth die . . . the wives of the said king . . . every one of them goe together to a place appointed, and the chiefe of the women hath a ball in her hand, and throweth it from her, and to the place where the ball resteth, thither they go all, and turne their faces to the eastward, and everyone with a dagger in their hand stab themselves to the heart."

There has been some controversy about this dramatic occurrence of the final conquest, which took place in 1905.

Personally I can see no reason to condemn the Dutch for firing on the female battalion. Surely a *kris* is a deadly weapon when wielded by a woman, as well as by a man.

The real cause for criticism might be found in the reason for this invasion. A Chinese junk was wrecked on the South Bali shore of the Indian Ocean. The Balinese helped themselves to the salvage, which seems to have been against the law of this land, already nominally under Dutch sway. The Chinese claimed that the junk was loaded with silver coin, but the Balinese denied it and refused indemnity. The Dutch expedition was sent to enforce the royal authority over the coast line. Surely any diplomat would have known better than this. To start such a war there should always be at least one insulted missionary.

The women of North Bali are likely to take off their shirts as soon as they get inside the walls of their own homes. In South Bali there is already a Government effort to make the women wear shirts in the towns, such as Den Pasar, but it is not strictly enforced.

Chapter II

Sir Thomas Stamford Raffles, the father of the British empire in Malaya and the patron saint of Singapore, was the dominant figure in the Dutch Indies during the interregnum of the British, and he visited Bali in 1815. In his monumental "History of Java" (London, 1817, Vol. II,

Appendix K) he gives an account of the people of Bali which still retains a remarkable degree of precision if you make allowances for his assumptions. Of course one must realize that this was a haughty soldier and governor, accustomed to the suave fawning of the long-conquered Javanese, who quite naturally was taken back by the don't-give-a-damn pride and self-sufficiency of the Balinese.

Raffles was overwhelmed with pleasure at the legal and administrative systems of the island, but it is interesting to note that he was not impressed with its art. There is room for disagreement about the art. Certainly it has not a great deal of appeal for those who confuse art with luxury.

At any rate, here is what Raffles had to say about the people themselves:

"The natives of Bali, although of the same original stock with the Javans, exhibit several striking differences, not only in their manners and the degrees of civilization they have attained, but in their features and bodily appearance. They are above the middle size of Asiatics, and exceed, both in stature and muscular power, either the Javan or the Malayu. Though professing a religion which in western India moulds the character of the Hindu into the most tame and implicit subserviency to rules and authority, and though living under the rod of despotism which they have put into the hands of their chiefs, they still possess much of the original boldness and self-willed hardihood of the savage state.

"Their general indifference to the oppression which they endure, their good humour and apparent satisfaction, together with their superior animation and energy, give to their countenances, naturally fairer and more expressive than those of the Javans, a higher cast of spirit, independence and manliness, than belongs to any of their neighbours. They are active and enterprising, and free from that listlessness and indolence which are observable in the inhabitants of Java. To a stranger their manners appear abrupt, unceremonious, coarse and repulsive; but on further acquaintance this becomes less perceptible, and their undisguised frankness commands reciprocal confidence and respect. Their women, in particular, who are here on a perfect equality with the men, and not required to perform many of the severe and degrading labours imposed upon them in Java, are frank and unreserved. In their domestic relations their manners are amicable, respectable and decorous. The female character, indeed, seems to have acquired among them more relative dignity and esteem than it could have been expected to attain where polygamy has been long established.

"The conduct of parents to their children is mild and gentle, and it is requited by unreserved docility and obedience. To their chiefs they shew a respectful deference: among themselves they stand on a footing of equality, and feeling no dependence pay little homage. The abject servitude of Asia has not proceeded further with them than necessary obedience to indispensable authority. Their prince is sacred in their eyes, and meets with unreserved obedience, but their minds are not broken down by numerous demands on their submission, nor are their manners polished by the habit of being frequently with superiors. An European or a native, therefore, who has been accustomed to the polite and elegant manners of the Javans, or with the general courtesy of the Malayus, is struck with the unceremonious, rude and uncivilized habits of the people of Bali.

"In the arts they are considerably behind the Javans, though they seem capable of advancing rapidly. They are happily not subject to a frame of government so calculated to repress their energies, or to waste their resources. They are now a rising people. Neither degraded by despotism nor enervated by habits of indolence or luxury, they perhaps promise fairer for a progress in civilization and good government than any of their neighbors.

"They are strangers to the vices of drunkenness, libertinism, and conjugal infidelity: their predominant passions are gaming and cockfighting. In these amusements, when at peace with the neighbouring states, all the vehemence and energy of their character and spirit is called forth and exhausted. Their energy, their modes of life, and their love of independence, render them formidable to the weaker states in their neighbourhood, and secure them against all attacks from any native power in the Indian archipelago. What they now are it is probable that the Javans once were, in national independence as well as in religious and political institutions."

Chapter III

Dutch sovereignty over Bali had been recognized by the powers for centuries, but not until 1846 did the Dutch be-

gin to move in their baggage. Then they came to Singaraja (Buleleng) with twenty-three ships of war manned
by 1,250 men and mounting 115 guns, and carrying 1,700
soldiers, of whom 400 were Europeans. With rifles, four
guns, and four mortars, the soldiers fought all day against
Balinese estimated by the Dutch at 50,000, armed with
spears and *krises*. The Dutch captured Singaraja with a
loss of thirteen dead and fifty-four wounded. The Balinese
left 400 dead on the field.

In April, 1848, because the Balinese had not lived up to
their treaty, a second expedition went to Bali, with 109
officers, 2,265 men, and twenty-four guns, backed by a
naval force of nine ships mounting seventy-two guns.
They attacked near Jagaraga and were repulsed with
"heavy losses," re-embarking and abandoning the expedition.

Then in 1849, with 5,000 rifle-bearing men and 3,000
coolies, with three batteries and six mortars, and a fleet
of sixty vessels with 300 marines, the Dutch set out to
settle this Balinese business. Peace negotiations fell
through in North Bali, and after two days of fighting on
April 14 and 15 with an army of 20,000 men the Raja
of Buleleng sued for peace. After some side expeditions
in the North, the Dutch came in May to Klung Kung in
South Bali, seat of the supreme Raja. Klung Kung was captured on May 24, but at 3 A. M. the Balinese rallied in
one mad, heroic attack, which failed. The Dutch losses
were slight. The Balinese left 1,000 dead and 800
wounded. Bali was conquered.

But it was not until the final expedition in 1905 to South

Bali when the women stood up before the gunfire, that Bali completely submitted.

CHAPTER IV

Ming plates are found everywhere in Bali. Roosevelt had stacks of them, and they were not worth a whoop. All had flaws or were lopsided. But occasionally would turn up a precious green Sung plate, which had outlasted seven centuries.

Apparently during the Ming dynasty the Chinese used the Indies as a dumping ground for their inferior chinaware. This they could easily do, for the Balinese do not glaze pottery or have much use for plates. They could be sold junk without knowing the difference.

The Balinese have, however, found an unusual use for plates. There are several temples, perhaps forty years old or so, which are decorated profusely with soup plates and saucers of obviously European design. These have been set in plaster on the temple walls.

The *gamelans* of Bali and Java are related also to the musical instruments of Siam and Cambodia, though I am told that the music of these continental countries is quite inferior. The civilizations of all these lands, of course, came from a common root; and kinship may be seen especially in the dancing.

How great a part irregularity plays in the charm of old churches may be seen by contrasting the prim and uninspiring exactitude of Antwerp Cathedral's interior with the irregularity of Notre Dame de Paris. The great arches of Notre Dame, for an extreme instance, are all askew, making the nave at least a yard longer on the north side than on the south. This irregularity of ancient churches, which may be accidental (the pillars of the Parthenon are not in a straight line either), has been adopted as a principle by Raymond Pitcairn in building at Bryn Athyn, just outside Philadelphia, a lovely Swedenborgian cathedral. Upon it he is lavishing a great passion for medieval architecture and with a large corps of artisans is reviving the individuality of the old guild crafts. He is, of course, also lavishing a good many of the plate glass millions.

Walter Spies tells an amusing instance of the tuning of gongs. While in Java he was commissioned to buy a set of four gongs for an orchestra in Vienna. He found a satisfactory set, but it was slightly off the required pitch. A gong expert said he could fix that, and sat down with the large gong, a bronze bowl more than a yard in diameter. With a sharp file he went around the edge, gouging until he had before him a big pile of bronze filings. Then he struck the gong. It shuddered, in a tone all scattered like a dishpan's.

Spies's heart sank. The gong was ruined; the set was broken; $300 was wasted.

Then the gong tuner took a little hammer, and began

tapping around the filed edge, smoothing it. Gradually, as he tapped, the scattered tone came together, focussed to a perfect pitch.

Chapter VI

The *kains* worn in Bengkel, the only garments except for head scarves, were wide lengths of cloth, wrapped about the body and belted at the waist. They were identical with the Javanese *sarong,* except that the *sarong* has its ends sewed together, to make a cylinder of cloth. Most of them were of *batik* design.

Batik is dyeing by the unique Javanese method, which is not practiced in Bali. A design is drawn on the cotton cloth, and all except a part of this, which is to be dyed with one colour, is overlaid with hot wax. Then the cloth is put in dye, and the waxed part is not coloured. The wax is then washed out with caustic soda, leaving the white cloth with part of its design dyed in one colour. The wax is then replaced, leaving still another part of the design exposed, to be dyed another colour. This process may be repeated any number of times. The wax cracks, and permits irregular threads of colour to run through the design.

Batiking is a fine art, carried on with an infinity of imaginative designs inherited from the days of Java's greatness. But there are two kinds of *batik*. There is *batik tulis,* upon which the native will insist if he is not too poor; for this the wax is put on by free-hand drawing, run through a tiny pipe after the manner of a fountain pen; and making it involves fine craftsmanship. Then there is *batik chap*, which is what tourists generally pay their exorbitant prices for, with the wax put on by copper stamps.

The most popular *kains* in Bengkel had fancy diagonal stripes of brown and black. There were peacocks too, of red and blue, with trailing vines behind them.

For special occasions the Balinese wear their own hand-woven *kains,* in charming plaids and elaborate checkered patterns. Unfortunately analine-dyed thread has sup-planted the old vegetable colours for weaving, for it is cheaper and easier to get. Then there are *kains* with de-signs of infinite elaboration woven in gold thread, and others on which gold leaf is glued in intricate pat-terns.

It is especially interesting to watch the weaving of those *kains* which seem to be embroidered in gold thread. Across the warp, which stretches from a yoke held across the weaver's back, is laid a large pile of slender round sticks, each of which has several loops of thread, picking up its own combination of warp threads. With amazing speed the weaver plucks from the pile of sticks the com-bination needed for each rapid shuttling of her bobbin. It is pretty much like playing a harp.

Not only are the Balinese not swimmers, they are not seafaring folk at all. This is an important factor in the country's long retention of its individuality. The Balinese does not travel. Around the shore line have settled a good many Mohammedans, a rather scurvy lot by comparison, who do almost all the boating.

Chapter VIII

Cockfighting originally was doubtless a religious ceremony. Its name, *lobeu getih,* means "spilling of blood." Until 1926, fights were held every day, and were taxed for the benefit of the local temple funds, but since then have been prohibited, except on occasional holidays. Although cockfighters are sometimes fined as much as $40, an immense sum, the fighting of course continues under cover. In the old days it was not at all uncommon for a man to lose all his property cockfighting and gaming with Chinese playing cards, and it still happens.

To those who revolt at the cruelty of the sport and consider it a savage barbarism, I can only comment that George Washington and Thomas Jefferson enjoyed cockfighting. We moderns are not so very far separated in time from many things of which we disapprove.

Cockfighting also is universal in the Philippines. Nearly every Filipino has his rooster. He takes it with him to work, and fondles it in his lunch period. Not only at dawn but at any time of day, especially at siesta time, you can hear cocks crowing in Manila; and Sunday, from sunrise to sunset, is a big day in the pits outside the city limits.

There now occurs to me one of the mysteries of the East. Fighting roosters seem to be spontaneously generated. I have seen some thousands of game cocks, each kept in careful celibacy by his owner. But never, that I can remember, have I seen a game hen. This is a mystery that I cannot explain.

CHAPTER IX

Hinduism in Bali is greatly different from that of British India, as I have read of it. There are the same castes—Brahmana, Satria (Anak Agung), Wisia (Gusti), and Sudra—but all rigidity has broken down. Paradoxically, while the class distinctions are recognized, the real life approaches social democracy. Though Sir Stamford Raffles reported that in 1815 there was a small class of tanners who were not allowed to live in towns, there are now no outcastes in Bali.

There are no begging holy men. There are no beggars at all among the Balinese, for that matter. There was one beggar in Bali when I was there, an old Mohammedan. Begging has not even been learned by the children along the tourist trail. This lack of the outstretched hand, in contrast with all the rest of the East, is the thing which always first impresses the new arrival.

Suttee, the burning of widows, formerly was practiced. (Cf. first note on Chapter I.) A century ago a Raja of Buleleng was followed to the pyre by no less than 72 women, but I do not believe this practice was a burning alive. Suttee is gone, discouraged by the Dutch. The last instance was in 1905 in the Balinese colony on Lombok, the next island. In South Bali the last time was in 1895.

The Balinese did not object to giving up a custom which in any case did not suit their natural temper. Widows now return to their families or marry again.

There is no child marriage in Bali. Though girls mature early, they do not generally wed until they are well over fourteen.

In Bali there is a great deal of common sense, and one finds none of those hideous manifestations attributed to religion in British India.

To try to unravel the pantheon, mingled as it is with the nature gods of aboriginal Bali, would be a terrific task for the most expert of religionists. The fierce-faced Durga, wife of Shiva, who also has a good-natured personification as Paravati, is undoubtedly a counterpart of Calcutta's Kali. All the gods are somewhat changed in character.

Orthodox Hinduism is fundamentally monotheistic, and the many gods are but props to the imagination of men unable to contemplate infinity, representing the various aspects of the Ultimate. It is interesting that in Bali this ethereal Being has itself been personified in Tintiya, obviously a primitive tribal god.

Chapter XII

If it seem that a great many persons in this book have the name Madé, that is nothing to the way it is in Bali. Every fourth person you meet, male or female, seems to be named Madé.

Ida Bagus Gidéh straightened out this matter of names for me one day. Names are just names; they have no meanings that can be translated. "Ida Bagus" is a surname that goes with certain classes of the Brahmana. Concerning other names, he said:

"Write my name down, Ida Bagus Gidéh. Now write: one, Brahmana wife; two, Satria wife; three, Wisia wife; four, Sudra wife."

I gasped. All through the East I had failed utterly to get orientals to answer hypothetical questions. Now Gidéh was stating a hypothetical case, outlining what his children would be named if he had a wife of each caste.

The Brahmana wife's children would be called Ida Bagus and given each an individual name, such as Gidéh.

The Satria wife's children would be called Ida Bagus Mura, and individual names. The Wisia wife's children would be Ida Bagus Compiang.

The children of the Sudra wife or wives would be named in the order of their birth, with "Ida Bagus" prefacing the names that are given to all Sudra children. These names are: Wyan, first-born; Madé, second-born; Nonga, third; Nioman, fourth; Ktot, fifth. The sixth child is named Wyan, and the rotation is started again. Each child is also given a name of his own, but the rotation-names are very commonly used.

Ida Bagus Gidéh, by the way, had not married a Brahmana. But his first wife, a Sudra, became a Brahmana by courtesy, so that her children might bear the names of full-bred Brahmanas and be eligible to inherit the full privileges of caste.

Rules don't mean much in Bali, and as soon as you have a rule you find an exception. Madé Réi, for instance, a Wisia, has the Sudra name Madé.

Chapter XIII

When Ida Bagus Gidéh said that the Old Lady "ate" *chandu,* of course he meant that she smoked it. At one time opium smoking was practiced in Bali to a considerable extent, but the Dutch put a stop to it, except with a few licensees whose habit was incurable. Of these the Old Lady was one.

Chapter XVI

In British India *babu* is a somewhat contemptuous term for an English-speaking Bengali clerk. In the Dutch Indies the word is commonly used to refer to a maid servant, though if we had been pedantic in our Malay we should probably have spoken of our *babu* as *kokki,* or cook.

Sirone, bright young woman that she is, was a great help to Roosevelt in making his Balinese motion picture drama, which is to be shown soon after this book is published. In it she plays the role of the villainous Jokorda's sister.

Chapter XVIII

The Dutch wife, that convenience which Plessen found in his bed, first puzzles the voyager at Singapore, then becomes his familiar bed-fellow in every hotel bed throughout the Indies. It is a hard, cylindrical pillow, or bolster, a yard or so long and a foot in diameter. It lies longitudinally in the bed, and its purpose is ventilation. Strangely it makes cool snuggling. You prop an arm and knee upon it, and it balances you nicely on your side, with plenty of air spaces between your limbs. It is a very comfortable device in tropic heat, in which of course you don't use any covers but the mosquito net draped over your bed. Old timers in these islands can't sleep without a Dutch wife.

Chapter XX

Sayres, in going without a hat, was demonstrating one of the strange phenomena of the tropics. The first thing the tourist generally does in the tropics is to buy an elephant hunter's helmet. And in British India and in parts of Africa it would be courting death to go in the sun without a helmet. But in the Malay archipelago, for some reason, a helmet is not necessary. There is a mysterious difference in the quality of the sun's rays. Of course it is advisable to have the head covered, but a light felt or straw hat is quite satisfactory.

Up in the Philippines it is the same. Of course some wear helmets, but when you see a lot of them in the Manila streets you know a tourist ship is in.

Chapter XXI

Guna-guna, magic, is much less used in Bali, I believe, than in Java, where native amorous reactions appear to be more understandable to westerners. Quite possibly this impression is due to the fact that in Java the whites and browns have been living together for centuries, and thus have had time and occasion to pierce the veil of mystery which always must exist between the races. It is significant that most of the *guna-guna* stories one hears have to do with contacts between the races. Many of the stories, doubtless, are due to the credulity and superstition of ignorant white folk.

It is a fact, however, that in Java the *dukun,* or witch-doctor, is a common and sometimes sinister figure. For the same black art which may be used for encouraging amour may be used also for revenge. One of the most interesting scandals in Java in recent years had to do with a half-caste family of immense wealth, headed by a tyrannical old woman. One of her granddaughters married a Dutch Lieutenant, who incurred the matriarch's wrath. Through efforts of a witch-doctor to blackmail her, it came out in the courts and papers that she had tried *guna-guna* against the young white man.

Guna-guna, in such cases, may mean poison, with which

the *dukun* reinforces the potency of his mutilated wax images and other charms. One hears in Java of lingering, mysterious illnesses and death, from unidentified poisons (powdered glass, perhaps, or bamboo) ; and the victims may be husbands or their new white wives, jealously hated by discarded mistresses.

Be that as it may, the love potions would seem to be less harmful or more amiable in their effects. There are those, however, who maintain that the potions are effective, by producing a lassitude and melancholia which easily succumb to feminine blandishments.

CHAPTER XXII

While I am about it, I can see no good reason for remaining silent on the subject of *sukyaki,* the culinary masterpiece of Japan.

You have a heavy iron pan, over a charcoal brazier. In this is put a lump of butter, or fat, some soya (suey) sauce, and a tablespoonful of sugar. Into this, when boiling well, are dropped very thin slices of raw beef and bits of Japanese onions, which are something like scallions of extremely delicate flavour. *Sukyaki* should be eaten with chopsticks, with the shoes off, sitting on the floor beside the grill. You have in your hand a cup in which is beaten up a raw egg. Picking the boiling *sukyaki* from the pan, you dip it in the egg to cool it, before you put it in your mouth. You have a dish of rice on the side.

It is important that everyone eating from the same pan

be of equal skill with chopsticks, or the inept ones will go hungry. There must be someone to replenish the pan continually while you eat, preferably a pretty kneeling girl in a kimono.

A friendly Japanese professor whom I met on the Kobe train confided that during his twenty years at Amherst his guests had often insisted on having *sukyaki,* and in the absence of Japanese onions he had found it satisfactory to use the shoots of ordinary onions, sprouted in the cellar.

There are those who make *sukyaki* of strong onions, cut the meat in cubes, introduce mushrooms and other extraneous matter, and spoon out the mixture over rice. Such heresy cannot be too highly condemned. Those who practise it are fit for nothing better than chop suey.

CHAPTER XXVII

The Malay opera, *Stambul* or *Bangsawan,* which the musicians of Belaloan identified with Mendelssohn, is a hilariously hybrid form of dramatic entertainment that tours east from Singapore. It invented a clownish Hamlet in modern dress, and with a telephone dated up Marguerite for Faust. There are no limits to its childish adaptations, and its tatterdemalion costumes are poems of incongruity. It has a western-style orchestra, and its songs are a mixture of native airs and the stickiest bits of sentimentality that drift through from the West. I was agonized one night at the City Opera in Singapore while the wicked Chinese Magistrate beat to death his beautiful

slave girl. It was neither her terrible fate nor her screech-
ing voice that troubled me; it was that as she died she sang
a tune once recorded on a cheap phonograph disk by harp,
flute, and violin, and I could not remember the confounded
thing's name. Not until the Captain in *Show Boat*, some
months later, struck it up on his fiddle did I remember that
it was the good old melodramatic standby "Flower Song,"
companion piece to "Hearts and Flowers." The slave
girl's soul sang it again in the next act, when led before a
"star king" in a world of icebergs, green, and purple lights.
The intermission between these two touching scenes was
taken up by two pretty half-castes who did a Charleston
well above the New York cabaret average. Such is the pitiful motley art of modern Malaya; and it finds little favour
among the musicians of Bali.

PHOTOGRAPHS BY ANDRÉ ROOSEVELT

Watery terraces, tinged green with rice

Fisher casting his net

*Ida Bagus Gideh's baby was all dressed up
for his birthday party*

Sironé and Nioman

Little Renkog would help Madé pound rice

In the market

A peanut stand

Day after day Renang was weaving

Janger Girl

Janger

Janger girls in every-day garb

The Janger ended in masked drama

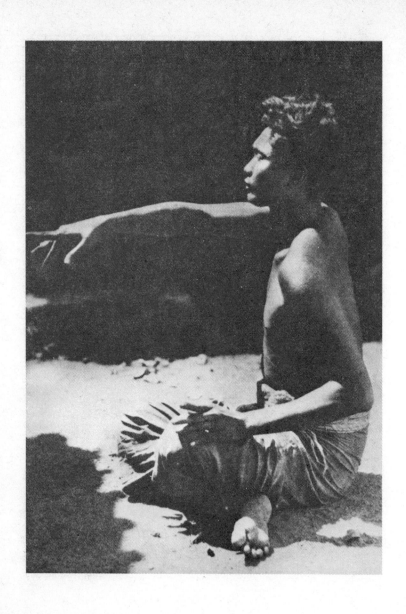

Ida Bagus Gideh in dancing pose

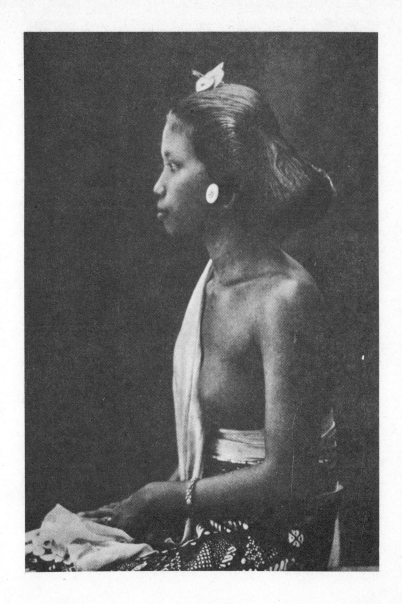

Sayu

Madé Rei and Runis dressed like other girls when not dancing

Madé Rei and Runis dance in golden costumes
before a gold-framed gong

In a Nirvana of sound

Mudari plays the joblag

Procession to a temple feast

Temple gate

Kumis put his fighting cocks, in their baskets, outside his gate

The hold-up and airplane panels flank the temple door
on the preceding page
Below: A traditional Ramayama bas-relief

Above: Terracotta Togogs from Bengkel
Below: Padanda Ratkarta's Drawing

A Palm leaf panel
Vishnu riding the garuda

Above: decoration from a musical instrument; grotesque
Below: Tintiya, God above all; a temple panel

Fierce-eyed Durga, with hand upraised

High caste man at prayers

Padanda Ratkarta rings his bell

All overgrown with moss

Carrying a cremation pylon

Cremation pylon at runway

Some other Oxford Paperbacks for readers interested in Central Asia, China and South-East Asia, past and present

Titles marked with an asterisk have restricted rights